The R.A.M.S. Library of
Alchemy

Volume 39

A Revelation of the Secret Spirit

Giovanni Battista Lambi

Also includes

Trifertes Sagani, or Immortal Dissolvent

Cleidophorus Mystagogus

and

Aphorisms of Urbigerus

R.A.M.S. Publishing Company

A Revelation of the Secret Spirit

Giovanni Battista Lambi

Also includes

Trifertes Sagani, or Immortal Dissolvent

Cleidophorus Mystagogus

and

Aphorisms of Urbigerus

Produced by

Restorers of Alchemical Manuscripts Society
1982

R.A.M.S. Publishing Company

R.A.M.S. Publishing Company
117 Rutherford Lane
Stuarts Draft VA 24477

A Revelation of the Secret Spirit

First Edition 2015

ISBN-13 **978-1511988872**
ISBN-10 **1511988878**

Image Processing by Philip N. Wheeler

Printed in the United States of America

Table of Contents

Disclaimer 7

Introduction 9

A REVELATION OF THE SECRET SPIRIT 11

The Preamble 24

The First Chapter. 27

The Second Chapter. 30

The Third Chapter. 37

The Fourth Chapter. 40

The Fifth Chapter 49

The Sixth Chapter 55

The Seventh Chapter. 59

The Eighth Chapter. 65

TRIFERTES SAGANI 73

AN EPISTLE TO THE READER 74

CHAPTER I 81

CHAPTER II 91

CHAPTER III 99

CHAPTER IV 112

CHAPTER V 133

Aphorisms of Urbigerus 143

A Word from the Publisher 175

Dedicated to Hans W. Nintzel,

American Alchemist

and

Founder of the

Restorers of Alchemical Manuscripts Society

(R.A.M.S.)

Disclaimer

Liability: The publisher does not warrant or assume any legal liability or responsibility for the accuracy, completeness, or usefulness of any information, apparatus, product, or process disclosed. The publisher makes no representation as to the accuracy or completeness of the contents of this book and specifically disclaims any implied warranty of merchantability or fitness for a particular purpose. No warranty may be created or extended by written sales materials or sales representatives. You should obtain professional consultation where appropriate. The publisher shall not be liable for any loss of profit or other commercial or personal damages, including but not limited to special, incidental, consequential, or other damages.

A Revelation of the Secret Spirit
by
Giovanni Battista Lambi

Introduction

Philip N. Wheeler

According to Adam McLean's provisional list of 1179 authors of alchemical books published before 1800, Giovanni Battista Lambi wrote and published only one book on Alchemy. The beginning of this book notes that it was written by an anonymous author in Latin. Mr. Lambi translated the text into Italian, and an unknown Englishman (R.N.E.) translated the work into English. It may be that the author's name is a pseudonym.

This volume of The R.A.M.S. Library of Alchemy also includes "Trifertes Sagani" by Cleidophorus Mystagogus (a pseudonym).

Finally, "Aphorisms of Urbigerus" is included. These 101 aphorisms were written by Baro (or Baru) Urbigerus, a 17th century writer on alchemy. These aphorisms claim to set out completely the theory of the alchemical work for the preparation of the Philosopher's Stone.

All three of these works were selected by Hans W. Nintzel for the R.A.M.S. Library.

A REVELATION OF THE SECRET SPIRIT

Declaring the most concealed Secret of ALCHYME.
Written first in Latin by an Unknown Author, but
explained in Italian by John Baptista Lambye,
Venetian.

Lately translated into English by R. N. E.
Gentleman.

Unto so high a Secret, who shall approach?

He brought Water out of the rock. Psalms 77. v. 13.

And only out of the hardest Stone. Deut. Chap. 32.
v. 19.

B. M. 8610-a a. 11.

MSS. Note.

The Translator (see the Epistle Dedicatory) was
evidently a native of Scotland. His Initials R. N.
E. might stand for Robert Napier Esq: (or of
Edinbrough?) a younger son of Napier of Marchiston.
London

Printed by John Haviland for Henrie Skelton, and are
to be sold at his shop a little within All—gate.
1623.

A REVELATION OF THE SECRET SPIRIT

Hermes, Plato, Aristotle, and other Philosophers in former times flourishing, the original Springs of Sciences, and the inventors of liberal Arts, earnestly approving the Vertues of things under the Heavens, did inquire with great desire, if anything was amongst the Creatures that might save mans body from all corruption, and preserve it alive for ever. Unto whom it was answered, that there was nothing that could deliver our corruptible body from death, but that there was one thing that could remove all corruptions, renew youth, and prolong short life, as in the first Patriarches; because unto the first Parents Adam and Eve, for penance of Sin death was given, which will never be separated, from the whole posterity. Wherefore the said Philosophers, and many others most painfully seeking that ONE THING amongst all things, have found that it which should preserve mans body from corruption, and prolong life, is such amongst qualities as the Heaven amongst Elements. They understood the Heaven to be above the Essence of the four Elements, and so that to be above the Essence of the four qualities.

The Heaven in comparison of Elements, is called Quintessence, because it is incorruptible, unchangeable, not receiving strange impressions; so also that thing, in respect of the qualities of our body, is incorruptible, expelling all superfluities from the body. It is called of the Philosophers by the foresaid name, neither hot and dry with the fire, nor cold and moist with the water, nor hot and moist with the air, nor cold and dry with the Earth, but of all the elementary qualities a perfect proportion, a true conjunction of natural power, a

special addition of Spiritual Virtue, and an
inseparable union of body and soul: A substance most
pure, most precious artificially extracted from an
incorruptible body, which no wales can be destroyed,
nor in anything be defiled with the Elements.
Whereof Aristotle did make an Apple, with whose
smell he did prolong his life, when through length
of age, he could not eat nor drink, fifteen days
before his death.

This Spiritual substance is that only thing, which
from above was shewed unto Adam, the most desired by
the holy Fathers, which Hermes and Aristotle do call
the truth without lye, the certain most certain, and
the secret of all secrets, hid from Nature, and the
marvellous final conclusion of all philosophical
Works.

In the which is found the dew of the Heaven, the fat
of the earth, and what the power of man cannot
expresse, in this Spirit is found; because as Morien
saith, who hath it, possesseth all Things, and shall
need the help of no body in anything, because in it
is all temporal felicity, corporall health, and
earthly properity.

This is the Spirit of the Quintessence, the Spring
of Sublunary health, upholder of Heaven, retainer of
Earth, Mover of Sea, stirrer of Winds, sender of
Rain, container of all things and virtues, and
spiritual and chosen above all subcelestial Spirits,
giving health and prosperity, Joy and peace; it
yieldeth love, dissolveth hatred, and generally
removeth all evils, most speedily cureth all
diseases; finally, destroyeth misery and poverty,
maketh and bringeth all good, cannOt speak or think

evil, giveth unto man what his heart desireth, unto the good temporal glory and long life, but perpetual pain unto the evil that use it.

This is the Spirit of Truth which the World cannot apprehend, but by the grace of inspiration, or the teaching of them that know it; it is of an unknown nature, wonderful virtue, and infinite power; This the Saints from the beginning of the World have wished to see. This Spirit, by Avicen, is called the Soul of the World, because as the Soul moveth the members, so this Spirit moveth all bodies; And as the Soul is in every part of the body, so in every elementary thing this Spirit is found; which is sought by many, but found by few, it is thought to be far off, and found near at hand, because in everything, place, and time, it is found having the virtues and effects of all things, and being equal in all the Elements, and whatsoever is proper to everything, in this only thing is found most effectually. By whose virtues Adam and the Patriarches had health of body, and length of life, and many others have flourished in riches.

Which being laboriously sought and carefully found, the Philosophers have hid in aenigmatick terms, that they should not shew so worthy a thing to the unworthy, not to throw so precious a pearl among Hogs, which if it were known to all, the study and labour of all men would cease, and man would desire no other thing but only it, and so men would live unworthily, and it would be the cause of the ruin of the whole world, as well through health as through abundance, men would much more offend God: Because the eye hath not seen, nor the ear hath not heard, nor it hath not hitherto entered into the heart of man, what the Heaven hath naturally placed into this Spirit.

Therefore I have briefly compiled certain properties of the said Spirit, approved by Philosophers, unto the praise of God and the profit of good men, that they might most devoutly magnify God in his gifts, at least they who hereafter shall receive such a gift, because these gifts are not belonging to everyone, but to them whose minds shall be good. Now what properties and virtues that Spirit hath in every essence, and how it appeareth corporally, that it may the sooner be found and known, hearken with the understanding of the heart.

In the first Essence, it appeareth in an earthly body foul and full of infirmities, in which it hath a property and virtue of curing wounds and corruptions in the intrails of men, it purge'th putrefaction and stink abiding in any place whatsoever, it cureth all things inwardly and outwardly.

In the second essence it appeareth unto the sight in a watery body, somewhat fairer than the first, containing corruptions, but more plentifully working his virtue, never unto the truth, and in every work more powerful, in which generally it giveth aid to all sickness both hot and cold, because it is of a hid nature, chiefly it helpeth them that suffer venome in the breathing parts, for it chaseth venom form the heart, dissolveth without violence things contained in the lungs, and (notwithstanding the commotion) it doth consoled the same ulcerated, it cleanseth blood, it purifieth corruption contained in the breathing parts, and it preserveth them cleaned from corruption; being Thrice a day drunk by any that languisheth, it maketh good hope and etc.

In the third Essence, it appeareth in an airy body oily, almost freed from all diseases, in which it sheweth wonderfull works; for it helpeth young men to last in body, state, strength, and beauty, if they use it by little and little, and in a small quantity in their meat, because it suffereth no ways melancholy to exceed, nor choller to burn. Moreover, above measure it encreaseth blood and seed, and therefore it behoveth them who use it, often to be let blood. Also this oil doth open the Nerves and Veins; and if any member be fading, it reduceth it to his measure; and if a young man before the state of age, hath an eye burst, if one drop be put therein every day, and that he be quiet for a month, without doubt his sight shall be restored. And if anything be putrified in any member, or superfluous, it dissolveth it speedily, and separateth it. And if it find it diminished, it restoreth it, & etc.

But in the fourth Essence, it appeareth in a fiery body not fully cured from all diseases; containing water, and not fully dried; in which it produceth many virtues: The old it maketh young, and if in the hour of the hickoake of death, there be given of this fire, so much as the Weight of one grain of Wheat tempered with Wine, so that it go down the throat, it reviveth and entreth, and warmeth, and pierceth even to the heart, and suddenly annihilateth all superfluous humours, and expelleth poison, and vivifieth the nature of the heat unto the Liver. And if old folk use this fire in a moderate quantity, and join thereto the water of gold, it removeth the infirmity of age; so that they may enjoy young hearts and bodies; and for this it is called the Elixir of life.

In the Fifth and Last Essence it appeareth in a body equally glorified, wanting all tanents, shining like the Sun and Moon. In which it hath all the foresaid Virtues and properties, which it possesseth in other essences, both fairer, and more wonderful. For his natural Works are esteemed the miracles of God, because if it be put to the roots of the bodies of trees long dead and dried, are made living, flourishing and fruitful; and if the lights of a Lamp be mixed with the self (same?) Spirit, they are not extinguished, but are burning eternally without diminishing.

And it maketh the precious Stones of Crystal most costly with divers colours; they that are of the mine shall never be better, and it doth many other things, which are not lawful, to be revealed unto the unjust, which are esteemed impossible unto man, because it cureth all bodies both dead and quick, without any other medicine.

By Christ Jesus witness I do not lye in anything, because the influences of all heavenly bodies, which in all and everything are infused, are found in it.

In this Essence it sheweth the treasures drowned in the Sea, and hid in the earth, and it maketh all the bodies of metals most pure Gold and Silver, and nothing like to it is contained under the Heaven.

This Spirit is the Mystery which was hid from ages revealed to some Saints, to whom it pleased God to make known the riches of Glory, which remaineth fiery in water, and carrieth with it earth in the air, and out of his belly floweth floods of living water and life.

This Spirit flies through the midst of the Heavens, as a morning cloud, containing burning fire water, and earth clarified in air. It expelleth the malice of Saturn and Mars, joining Jupiter with the Moon and Mercury, and in the light of the Sun, giving unto his Sister Venus honey of the rock, and liveth with her forever.

And albeit these works appear erroneous and false unto the Readers, yet to the skillful and those that prove them actually, they are true and possible, if the figurative speech be faithfully understood, therefore unless thou understand sufficiently, do not intrude thyself any ways into this spirit, because God is marvelous in his Works, and there is no number of his Wisdom. This Spirit in a fiery nature is called Sandarasha, in airy Alkahest, in watery Azoch, in earthly Alcochaph; (?) by which means they who seek him are deceived, thinking the spirit of Life to be in such things, which to our knowledge be of no value.

And albeit this Spirit whom we seek, by reason of his property is called by these names; yet in these bodies, he is not, nor cannot be; for the glorified Spirit cannot appear but in a bodie agreeing to his Kind, albeit he is named by these and many other names.

Neither should any man think, that there be divers
spirits, but howsoever it is called, it is one and
the selfsame spirit, that worketh all in all things.
This is the spirit whom in ascending the clearness
of the Heaven hath overshined, and in descending the
puritie of the earth hath incorporated, and flying
above the wideness of the Sea hath received.
It is not of the lower Hierarchie, where is RAPHAEL
called the Angel of God, most subtil, most precious
and most pure, unto whom as unto a King all the rest
obey.

This Spiritual Substance is not celestial, nor
infernal, but a certain airy body gloriously
purified betwixt the highest and lowest, placed in
the midst, spiritually animated, wanting reason but
fruitfully profiting; above all things under the
Heavens choised and adorned.

This divine work is made too profound, that the fool
may not understand it, because it is the last of the
secrets of Nature.

This is the Spirit of the truth of the Lord, who
hath replenished the Globe of the Earth, and in the
beginning was carried upon the waters. Whom the
world cannot conceive, but by the grace of the
inspiration, or the teaching of those that know it,
and whom the whole world hath desired, for his
virtues that appear inestimable.

For it entreth the Planets, chaseth away the Clouds,
giveth clearness to everyone, and converteth all
unto Sun and Moon; it giveth all health and
abundance of Treasure, it cleanseth the leprosie,
cleareth the sight, comforteth the sad, healeth the

sick, rendereth hid treasures, and generally cureth all discord.

By this Spirit the Philosophers have found out the seven Sciences, and had abundance of riches. By this Moses made the Vessels of pure Gold in the Temple, and King Soloman many and precious ornaments to the worshipping of God.

And many others have made wonderful and great workers, who built the Ark, Moses the Tabernacle, Solomon the Temple, Esdras recovered the Law, Mary the Sister of Moses kept hospitality, Abraham, Isaac, Jacob, and other godly personages obtained length of days with abundance of riches, and flourished, and the godly knowing it glorified God. Therefore the obtaining thereof is better than the traffick of Gold and Silver, because it is more precious than any works; and all things that are desired in this age cannot be compared unto it, because it is proved and found perfect and infallible.
For in it only consisteth the truth, wherefore it is called the Stone, or spirit of truth, and in his works there is no vanity, whose praises I cannot express because I am not sufficient to tell his Virtues.
For his goodness, property and vertue, is greater than the mind of man can conceive, or the Tongue express by words; because the properties of all things are hid in it, and all that nature hath given to other things, in it being true, is truly found; What shall I say more? There is not, was, or ever shall be, any who shall

8.

search nature deeper. O the height of the Wisdom of God, because what all bodies have, thou hast enclosed in the power of one SpirIt! O ineffable glory, O! inestimable Joy, shown unto mortal man! Because the corruptible things of nature by virtue of the Spirit are made better. O Secret of all Secrets, health and remedy of all, the last search of nature under the Heaven, and the wonderful conclusion of the ancient fathers, and of the latter wise men, and of all Philosophers, the which the world and all the earth desireth. O most wonderful and much praise-worthy Spirit. It is the purity in which all delights and riches are contained, and also the fruitfulness of life, Science the strength of Sciences, giving temporal joy to those that know it. O Knowledge worthy to be desired and beloved above all things under the Moon by which nature is strengthened, the hearts with the members rejoiced, flourishing youth preserved, age removed, infirmitie destroyed, and most pleasant health kept, abundance of goods had, and all that delighteth man plentifully purchased.

O Spiritual Substance commendable above all Things! O superiour virtue in Things invincible! Which albeit it hath appeared contemptible unto the unwise, yet to those that know it, it is to be beloved, for praise, glory and honour; because it expelleth naturally all manner of death caused by humours: O Treasure of Treasures! O Secret of Secrets! This is the infallible Substance called and named by Avicen the soul of the World, most pure, most perfect, and most powerful. Nothing under the Heavens so precious, of hid nature and of wonderful virtues, operation, and infinite power. Unto which nothing is like amongst creatures, which hath all the virtues of the bodies under the Heaven, for out of it floweth waters of life, honey and oil of everlasting health, and so with the rock, arid honey he hath filled them. Therefore saith Morien, "Who hath it possesseth all things, and needeth no ways the help of others."

Blessed art thou, O Lord God our Father, who gave this Knowledge and understanding to the Prophets and Philosophers, that so they have hid it, that the blind filled with worldly lust might not find it, and the well-disposed by this have praised thee.

Grant that it may be discovered to none, but to the lover of thee, and to the desirer to do good things by it. Because who unworthily discovereth or revealeth the Secret of this thing, he is the breaker of the Heavenly Seal, and the hid revelation, so far as in him lieth he diminisheth the Majesty of God, and he is near unto many miseries to follow him.

And therefore with a godly heart I beseech all you faithful in Christ, having this knowledge, that you would not speak thereof nor communicate to any but to the godly livers, and disposers of themselves virtuously, long known and proved, and that you praise God who hath given such a treasure unto men.

This many do seek, but few do find it, for the defiled with vices or polluted, are unworthy to know such things. Therefore it is not shewn but to the devout, because it is incomparable to all pieces.

God being my witness, I do not lie in anything, albeit it appear impossible unto the foolish. For

none is, was, nor shall be, who hath so much
searched the depth of nature. Blessed be the most
High and Almighty God, who hath created this
Science, and hath been pleased to show unto the
faithful the Knowledge thereof. Amen.

So endeth this most worthy, and most excellent work,
the Work called the Revelation of the Secret Spirit,
in which all the secrets and mysteries of this World
are hid, & etc.

But the Spirit a power, is one marvellous and holy,
having for a gift the whole world, for it containeth
it in itself, is, shall be, and was also a Fifth
Substance.

The Preamble

In the name of God then to begin this business, I say, that this Philosopher would have shewed and declared the wonderful Virtues and properties of that Secret Spirit; saying, that it is such a Thing, that removeth all corruption, in these words: "But that there was one Thing that could remove all corruptions. And after he saith, "That the said Philosophers painfully seeking that one thing amongst all things, & etc." The which a little after he saith, "To be a glorious Spirit, called Quintessence."

But he telleth not, out of what thing it may be taken; unless that he saith, "It is such a substance most pure, most precious, and most subtil, from an incorruptible body artificially extracted, that no ways can be destroyed, nor in anything defiled with the Elements."

Agreeing with John de Rupescissa, who speaketh thus, "And I said that the most High created the Quintessence, which is extracted out of the body of Nature created by God, with humane Artifice, & etc."

Much less doth he show the manner nor the mastery of being able to have it, except that in some place he toucheth a little, and that under metaphores and Figures; therefore to declare this Text, I will ground myself upon three principal parts, to the end that this most noble Spirit may be found.

The first part shall be upon that body whereoutof the said Spirit may be taken. The second shall be, where he saith, "How that Spirit corporally appeareth that it may be the sooner found and Known, & etc."

And because the Author followeth five apparitions, continuing without making other distinctions, yet because the matter doth require it, I will divide it into two parts, and for the second part shall be taken the four apparitions.

And for the third part, that fifth apparition where he saith, "In the fifth and last Essence it appeareth in a glorified body, and etc."

In the first part shall be shewn, that so as the text saith this Spirit is ONE THING AMONGST ALL THINGS, that as yet the said Spirit or Quintessence in not found, but in one only Thing created by God; according to the Authority of all the Philosophers, who have written of this divine Science; who all do affirm, that there is nothing but one only Thing, of the which, and by the which, and with the which, the desired end is obtained. And in the second shall be seen, that the apparition of the said Spirit in four corporeal forms, is no other Thing, but the separation of the four Elements.

And in the third and last part shall be shewed, what is to be understood by the conjunction, and union of the said elements, after their perfect rectification, reducing them in a glorious fifth being, called Quintessence or Elixir, and in many other divers names nominated by divers philosophers, and it shall be that, which this philosopher understandeth, when he saith, "But there is one Thing, & etc."

Where to follow this order, for the better declaration of the Text before alledged, the first part shall be divided into three parts.

In the first shall be shown (as is above said) that there is one only thing, in which the Said Spirit is found.

In the second shall be seen (by the means of many authorities and sentences of divers philosophers) if it may be judged, what Thing is this only Thing.

And because there is no comming to the separation of the Elements, if first that body; or truly one thing (as they say) be not disposed, that it be fit for the said separation, which is nothing else, but the reducing it to his first matter, and this shall be taken for the third part, in which shall be seen the necessity of the said separation. Thereafter in the second part of the Text, shall be showed the manner of the separation of the Elements described by an uncertain Author, inserting therein some fair glosses of other philosophers for the more declaration of the Mystery.

And in the third and last part shall be seen the conjunction of the said Elements, of which shall grow a Crystalline and glorious Stone, called ELIXIR, or Quintessence, (as some would have it) and it shall be (as is said before) the secret Spirit of our Philosopher, & etc.

The First Chapter.

Wherein it is proved that there is only one thing, out of which the Secret Spirit, or the Philosophers Stone, may be taken.

Hermes, Father of all Philosophers saith, that "Our Stone is made of one Thing, because all our Science and mystery is made of our Water, that is Copper."

And Aristotle. "In one Thing only consisteth truth, and in multitude vanity."

And Pythagoras. In the book of his documents speaketh; "And I say unto you, that the foundation of this Art, for which many have perished, is one Thing, that the Philosophers say is stronger and higher than the natures, and to the foolish is baser than anything we esteem."

And Geber. "It is one Thing, one Medicine, to which we add nothing, we diminish nothing, except that in separation we remove the superfluous."

And Rasis: "Know my Son, that it is one disposition, and one nature, and one work, and one vessel to make the white red."

And Morienus; "This mystery is wont to be made of one only thing, therefore put this Theory in thy Mind, for thou needest not many things, but only one Thing."

And Lucas, in Turba; "For one verity is one, as which is the Spirit that we search."

And another uncertain Philosopher saith; "For our matter is one, containing in itself the essence of all seven Metals, and in that substance is the living Spirit which we seek."

And in the book entitled, "The Secret of Avicen;" it is written thus, "And know for certain, that the Philosophers cared not for the names, but one name and one action; To wit, to seeth the Stone, and bring forth his Soul, because their Stone is always one."

And John of Damascus, in his Rosary of Phoebus, saith; "Therefore it is doubted of the Stone, which is called the Philosophers, which is it, and what; seeing it was never named openly by any Philosopher: Therefore in this many judged divers Things, when in one only consisteth truth. We do hold him dearly, Teaching to avoid all others, and surely it is manifest by the Philosophers Writings, that it is one Thing, and that no strange thing should be joined thereto, because nothing agreeth to a thing, but what is neerer to his own nature."
And Arnoldus de Villanova, in his Rosary saith; "Therefore it behoveth the searcher of this Science to be of a constant will in work, that he presume not to try sometime this sometime that, because our Art is not perfected in the multitude of things, for it is one."

And Raymund Lullius in the 49 Chapter of the Theoricke of his Testament saith; "Thou must no ways be ignorant hereof, seeing it is necessary, that our medicine of one only kind and one sole nature be made." And in the Seventy fifth Chapter; "Moreover we say recapitulating, that seeing this thing is of one sole nature, and of that only this is made which mighty men desire to have and search, and in many other places he mentioneth, which for shortness sake shall be left out."

And George Ripley, in his Ladder of twelve Degrees, saith; "Yet the matter of this work according to all the Authenticable Philosophers, is one only thing, containing in itself all necessaries to the accomplishing of its own perfection."

And Henricus Cornelius Agrippa in the second Book of his hid Philosophy, in the fourth chapter, saith; "There is one Thing by God created, the subject of all Wonderfulness, which is in earth and in Heaven, it is actually Animal, Vegetable, and Mineral, found everywhere, known by few by none expressed in his proper name, but covered in numbers, figures, and riddles, without which neither Alchemy nor nature or Magick can attain their perfect end."

And in the Rosary of the Philosophers it is written, "But I advise, that no man intrude himself into this Science to search, except he know the beginning of true nature and her government, which being known, he needeth not many things, but one Thing; nor requireth great charges, because the Stone is one, the medicine one, the Vessel one, the government one, and the disposition one & etc."

Many other Authorities and sentences of divers Philosophers (for the confirmation of this passage) I could produce, but for shortness sake, as also because I think that the foresaid sentences of so many worthy Authors, are sufficient enough for confirmation of this matter, we let them alone.

The Second Chapter.

In which shall be seen, (by means of many Sentences of divers Philosophers) if it can be judged, what thing is this only Thing.

This is the passage which abaseth the wings of the ignorant searchers of this glorious and divine Science. Ignorant call I all those (to wit of this Science) who are ignorant of the true matter, of which the Philosophers Stone is made, albeit that in other Sciences they were most excellent and most learned.

But before I proceed further, two causes do make me stand somewhat doubtful of the manner which I should hold in my writing. The one is two Sentences, one of Aristotle, and another of Geber.

The first of Aristotle in his light of light, saith, "That the Ancient philosophers have therefore so carefully hid this mystery, that few might understand it; because if it were vulgarly known, there were no further place for prudence: Seeing the foolosh would be equal to the Wise." And the other Geber: "Wherefore the Science is not delivered without interruption, because the wicked as well as the good, would usurpe it unworthily."

The other Cause is for the difficulty of the matter; but considering that this secret is the gift of the most high God, as all the Philosophers say, and chiefly "Morien", with these words: "And know you that this masterie and secret of Secrets of the most high and Great God."

Therefore I hope that our Lord Jesus will put such
form to my speech, that without offending his most
high Majestie, I may help the Children of the truth.
And if about the difficulty, and depth of the
matter, my forces being not sufficient, the mind
nevertheless is most ready, and most desirous, to
make the vertuous spirits professours of this noble
Science, participant of a part of my long trouble
and study made therein, nor yet as one that would
persuade myself to be such, that I understood the
secret hereof, but as a loving professor of the
search of it, wherein I have wearied myself for the
time of twenty-five years.

Now to return to our first purpose: I say then for
to begin this second part, that in the beginning of
the book called "The Blast of the Trumpet," it is
written thus: "Of the first essence, the first
Philosopher Thales the Milesian saith, God is the
most ancient of Things unbegotten, eternal; and
therefore Pythagoras saith: "I say that God was
before all Things, nothing was with him when he was,
and understand that God when he was alone in the
beginning Created One Substance, which he called the
first Matter; and of that substance he created the
other four, fire, air, water and Earth; of which now
being created, he created all Things, as well high
as low, and so before all other Things, except the
first matter, he created the four elements; out of
which he created afterward what he would, to wit,
divers natures, & etc."

And Raymund Lully, in the Third Chapter of the
Theorjcle of his Testament saith: "God created that
nature of nothing into one pure substance, which we
call Quintessence, in which whole nature is

comprehended & etc." And in many other places he saith the like; because he considereth, that this Science proceedeth from God, as all the Philosophers do affirm, Therefore Mireris saith: "That this Stone proceedeth from the most glorious high place, and of the greatest Creator, which hath put to death many wise men, which is like unto everything, whose manner cannot be spoken."

Where I considering the height and difficulty of this matter, as I have above said; seeing that out of the Philosophers the construction concerning this Thing cannot be had: I purpose to see if out of the holy new or old Testament some Juice may be extracted, studying the which in my Judgement, many most excellent sentences may be taken -to the purpose of this matter, and of the whole science, the which shall be left out, and only I will serve myself with that which seemeth unto me most to the purpose of this passage.

I find that the first thing that our Lord God created was the Light, as it appeareth in the first of Genesis: Thereafter he made that wonderful separation of the elements, whereby there comineth in my mind some sentences of Vincentius in his Natural Mirror; in the second book, the three and thirtieth Chapter; where he saith, "Therefore his Spheres, which is true light, and are begun from light, and in light all Things are accomplished, & etc." And furthermore, "From the light, he began, that he might show his works to be the works of light, not of darkness & etc. And after he saith, So also by his example, he that taught man to work in the light."

And in the 35th. he saith, "Therefore the first substance is light, & etc. and after following he saith, Everything therefore, how much it hath of light, so much it holdeth of divinity; because God is light, and everything having more of light than another, is called more noble than it: For in all Things, nobility is remarked according to the greater nearness, and partaking of divine essence. And this also is manifest by induction in the first bodies, when they are compared together, the water is nobler than the earth, because it hath more light than the Earth: Likewise the Air than the Water, and the fire than the Air, and the fifth body than all other: Therefore it is called amongst them the first and most noble; therefore the perfection of all these things which are in every order of the World, is light."

And in the 38th Chapter: he saith, "Therefore worthily amongst all bodies, the light holdeth -the first place. For as S. Augustine saith; "Every substance common to two substances, according to nature is before them both; but light is a substance common to fire, and Stars, which precede all other bodies. Wherefore the first of all bodies is light." After in the 39th Chapter he saith; "But light is caused in the air, not from the Air itself, or the form of the Air, but from the Sun, & etc." And he speaketh many other notable sentences, which shall be left to be studied by studious men. Now seeing the light is the first thing which God created, I may say unto you that the self-nature is derived from that light, as all Philosophers do affirm, saying, use venerable nature: And for confirmation hereof, I will see if we can agree, many pretty sentences of divers Philosophers, who speak of this Science in favour of this opinion.

But before we come thus far, I would know by what occasion, many and divers wise Philosophers have entitled many of their books belonging to this Science, under the name of Light; As Aristotle called one, The Light of Light. Rasis, five, to wit, The Light of Lights, The book of the Sun, The book of the Moon, The book of Clearness, The book of Light. John of Damascus, the Rosary of the Sun. John of Vieu, The Mirror of the Elements. Arnold de Villa Nova, The new light. Roger Bacon, The Mirror of Alchemy,[1] John de Rupeacissa, The book of lights; and many and divers others the like, which truly I cannot think that they would do it, but because this divine Science is the work of Light.

But perhaps some may say, the Philosophers use to say, when they will declare anything obscure, that they give light, or illumination. But I myself do not find that they have entitled their books of other Sciences under the title of Light, as they have done this. Let everyone believe as most pleaseth him.

Now let us come to the foresaid sentences, and first, Aristotle in his first book of the Secret of Secrets saith, "For with thee is the Light of Lights, and for this all darkness shall file from thee."

And Mirrors, in his book of documents, speaking of that secret Spirit, under the name of water, he saith, "And know that the Philosophers declare, that

---

[1] The R.A.M.S. Library of Alchemy Volume 32. -PNW

the permanent water is taken out of Light; but the light maketh fires, and the light shining, and transparent, becometh like one straying seeking lodging; but when light is conjoined unto light, it rejoyceth, because it came out of it, and is converted unto it, & etc."

And Albertus Magnus, in the preface of his right path, invocating the Lord God, prayeth and beseecheth him in this manner: "Thence in the beginning of my speech, I call for his help, who is the fountain and Spring of all good, that he through his pity and bounty, would vouchsafe to supply the smallness of my Science, that by the grace of the Holy Spirit, I may make manifest in my doctrine, the light which shineth in darkness."

And Raymund Lully in the 7th Chapter, in the Theorick of his Testament, saith; "Therefore son I say unto you, Take a Mine of the kind spoken of, in which are the two starred lights, which cease not to shine upon the earth, and they are the Sun and Moon who by their beams darken the fire."

And in the 48th Chapter: "Son when thou wilt work and begin our mastery, beware that thou work not but upon natures, or matters lightsome, and not upon others, whereof the lesser world is created.

And in the 10th Chapter of the Practice he saith, "Son it behoveth thee now to dissolve the light of the world, or a part of it, by the first regiment, etc." And in many other places he speaketh which shall be omitted, that I be not too long.

And George Ripley in the Chapter of the sixth degree of his Gates saith, "Therefore our stone is that starred Sun, budding like the flowers of the Spring, from whom proceedeth by alteration every colour, etc." And for shortness sake, I will put and end to this second Chapter; with a conclusion of an uncertain English Author, saying, "For indeed to speak without fiction, there is no other -to be sought but a body of the body, and a light of the Light." Which is as much worth as it were to say; "Separate the shadow from the beam, that is, from the Sun his earthliness."

The Third Chapter.

Wherein is proved, that of necessity it behoveth to reduce the body to the first matter, that it may be disposed for the Separation of the Elements.

Hitherto it is seen with the confirmation of all the Philosophers, that it is one Thing only, out of the which is taken the Secret Spirit, and with the which is made the Philosophers Stone: And furthermore are shewed many pretty sentences of divers Authors, by the means whereof it shall be left to be judged by the children of the truth, what thing can be this sole thing, or one Thing. Now in this Chapter shall be shewn (by authority of many authors) that it is necessary to dispose this Thing or body, that it may be fit for the separation of the Elements which could not be done, if first it were not reduced to the first matter: For anybody standing in his being hard, solid, and compact, is not fit for the separation of the Elements, much less for the metallick transmutation.

Therefore it is need (as above said) to reduce it to the first matter, according to the speech of Aristotle, in the fourth of the Meteors, who saith, "Let the Artificers of Alchemy Know, that the kinds of metals cannot be transmuted, unless they be reduced to the first matter, but the reduction to the first matter is easie, as Arnold de Villa Nova saith, and John of Vien in his Mirror of the Elements, and so affirm all the other Philosophers. And in the practise of Philosophers it is written: In the name of God, Amen, and with his help, let us reduce first the bodies into no bodies, until the natures be made thin, because in the beginning, the body is made a thin water, that is Mercury, etc. And in the Rosary of the Sun it is written, "Therefore every body is an Element, or compound of Elements, but the generation of any compound of the four Elements, consisteth of Simples. Wherefore necessarily it behoveth that our stone be reduced to the first matter, or spring of his Sulphur and Mercury, that it might be divided in the Elements, otherwise it cannot be purified nor conjoined etc."

And Villa Nova so saith; "For the first work of the Philosophers, is to dissolve the Stone into his Mercury, that it may be reduced to his first matter. And Raymund Lully, in the Seventy-fifth Chapter of the Theorick of his Testament saith: "But this division cannot be made by the change of nature, without loss of the property and the loss of the property cannot be made except that nature which is in an hard mass, and dry, with all her parts be turned in the likeness of that first nature, in which the age was first ordered by divine power, like unto a confused form, in which all middle things were confused, without the which nature could not accomplish her actions, etc."

Agreeing with Hermes, who saith: "All things were from one confused cold, or mass confused, by the mediation of One; that is, the creation of one Omnipotent God, and so all things were sprung from this thing, that is, all metals are engendered of our Stone, that is Quick-silver, as all things were sprung from this confused mass, and purged with one fitting, that is only by the command of God and his Miracle. So our Stone is sprung, and commeth out of a clayish mass, that is, Quick-silver changed, containing in itself the four Elements, which are Fire, Air, Water, and Earth, that is, heat, moisture, coldness and driness & etc."

And in the Rosary of the Philosophers: "The reduction of the bodies to the first matter Quick-silver, is no other than the resolutions of the congealed matter, by which the work is opened, by the centering of one nature into another. Whereupon the Philosopher said that the Sun is no other thing but ripe Quick-silver."

And this proposition or sentence of Aristotle, of the necessity of the reduction to the first matter, being so famous, I will not enlarge myself otherwise in alleging other Authors; but only I will labour in the next Chapter, to see if it be possible to know (by the means of many authorities of divers philosophers) what thing is this first matter.

The Fourth Chapter.

Where it shall be seen if it be possible, to know what Thing is THIS FIRST MATTER.

Having seen that the reduction to the first Matter, is necessary, that the matter may be fit for the separation of the Elements, now it is to be seen what thing is this first matter.

And it is above said, that all philosophers agreeing, do affirm this sentence, USE VENERABLE NATURE. Therefore in confirmation of this passage, I will serve myself only with some of their sentences, which shall seem unto me most to the purpose for the declaration of this matter.

Arnold de Villa Nova, in the first Chapter of his Rosary saith: "It is therefore manifest, that the operation of the Medicine is the operation of nature, and that the medicine itself is the same nature; for the medicine itself only is composed of nature, etc."

And Raymond Lully in the 72nd. Chapter of the Theorick of his Testament saith: "Our Mastery is by nature, and with nature, and of nature, and is made by means of nature.

And in the 76th. Chapter: "Wherefore who will make anything, let him make it by nature; and let him know the concordance of nature; because without that, never anything can be done. Seeing that nothing of the world which is facible, is beyond the

limited bounds or ways of nature, because by it and
with it, it is made and is to be made."

And in the 14th. Chapter of the Practick he saith:
"Son, if thou understand this, thou shalt understnad
and know how all things of the world are made by
nature, and how thou may make them to the respect of
nature, if thou can have the air which is caused by
nature, etc."

And in many other places he maketh mention, and the
Rosary of the Philosophers saith, "Whereby first we
make known, that all workers beyond nature are
deceivers, and that they work in a Thing unfit." And
therefore he saith: "In the Art of our mastery,
nothing is hid by the Philosophers, except the
secret of the Art, which is not lawful for any man
to reveal: Which if it were done, he should be
cursed, and should incur the indignation of the
Lord, and should die of an Apoplexy, wherefore all
errour in the Art ariseth of that, that they take
not the due Matter: Therefore use Venerable nature,
because of it, by it, and in it, our Art is
engendered, and in no other. Arid therefore our
Mastery is a work of nature, and not of the work of
man, and so who knoweth not the beginning, doth not
obtain the end, and who knoweth not what he seeketh,
shall not know what he shall find."

But because upon this Authority some may say, that
this philosopher intendeth and speaketh of the true
matter, on which we must lay the foundation; I say
that it is true, but out of what matter it behoveth
us to take the same nature as yet he maketh no
mention, which is inclosed in the center of the same
matter, witnessing Raymund in his 7th. Chapter of

the Theorick of the Testament, where he saith: "And we have said above, that in the centre of the earth is the Virgin earth, and the true Element, and that it is created by nature, therefore nature is in the Centre of everything."

Now having above shewed the necessity to reduce the body unto the first matter, and in this chapter proved that it behoveth to work with nature, so that it appeareth almost that this first matter is the selfsame nature, by the means of the authorities above alledged of the Philosophers, therefore for better declaration it is good to proceed from degree to degree.

First it is said the matter to be one sole Thing, thereafter we have spoken of the reduction unto the first matter; and now it appeareth that this first matter is (as said is) the same nature. It remaineth then to be seen, what thing is this nature, and it shall be the last conclusion about this passage. I say that of divers philosophers it hath been named with divers and infinite names some do call it Chaos, some Hyle, others the first matter, others a confused Mass, Matter without form, Confused Cloud, others Mercury; alledging that speech of Hermes, who saith: "In Mercury is all which the philosophers seek, etc."

And with many other infinite names, as I have above said, the which would be too much to desire to remember all. But I myself am disposed, leaving all other names, to name it under the name of Salt in this my little Treatise, alleging for confirmation of my opinion, a number of sentences of divers Philosophers.

And first we shall begin with Geber, who in his
Testament speaketh of no other, but of the Salt of
Metals, and sheweth that therewith is made the
Elixir, as he may see who would study it.
And the Rosary of the Philosophers saith: "The Salt
of Metals is the Philosophers Stone."

And a little further, "The Ancient Wise men, because
they spoke many things, did conclude upon Salt which
they call the Soap of the wise, and the little Key
which closeth and openeth, and again shutteth, and
no man openeth; without which little Key, they say
none in this age, can attain to the perfection of
this Science, that is, unless he know to calcine
Salt after his preparation & etc."

And after he saith: "Who hath not tasted the taste
of Salts shall never obtain his wish."
And Gratianus saith, "Of every Thing may be made
Ashes, and of that Ashes may be made Salt, and of
that Salt is made Water, and of that is made
Mercurie and of that Mercurie through divers operations is made Sol."

And Avicen saith: "Son if thou wilt be rich, prepare
Salts until they be a pure water, because Salts are
converted into a Spirit by Fire."

And Raymund in the 72nd. Chapter of the Practiek of
his Testament saith, "And we say, unto thee that the
said natures are nothing but sharp salts, etc." And
thereafter, "But Son we speak to thee with
revelation, that thou remember of the salt, which we
have told thee about, with one part of his
properties; because at no time must you understand
here of other salt; unless it be of Metals, which

are resolved into it, as by artifice you may see to the eye.

If therefore thou know how to sweeten this Salt, it will enter in the bodies, as the true nature which will stand inwardly, and can transform them from one kind to another; because seeing Salts are the first nature of Metals, of a Metallick propertie, by the friendship of that Thing, they are conjoined together. Seeing Salt is nothing but fire, nor fire is but Sulphir, nor Sulphur is but Quicksilver reduced into a Stone; after that it is the matter of a nature altered and changed from baseness to nobility."

Here clearly is verified and confirmed that passage above said: "That in the centre of the earth is the virgin earth, & etc., and that nature is in the centre of everything, & etc."

And in many other places mention is made, which for shortness sake shall be left out: O what labours, what sweats, what troubles, must be done What most thick and most hard walls must be broken and passed. And what ports and locks must be opened, before it can be penetrated and entred into that centre, where that blessed Virgin earth is found, otherwise by the said Raymund called the earth of labour. Arid truly the earth of labour it may be called, because it is purchased with great trouble and watching.

The which was well understood by Chrysogonus Polydorus, in his preface of Geber, when he said: "The Golden fleece is not given unto Jason, unless first he undergo the sure and dangerous labours."

And so much the more because it is to be known, that where the glorious God hath put great virtue there yet hath he left greatest difficulty for custody: But let us leave this, and return to our first discourse. I say then that I have gathered together many sentences of divers Philosophers, all which have treated of Salt.

Whose names to be shorter I will conceale, as also because some of these sentences are taken out of books of uncertain Authors. And I will repeat only their sayings with a continual order one after another.

And first the first saith, "Our Stone is no other Thing but Salt; who worketh in this Art without Salt, is like unto him who will shoot, not having a string. If the omnipotent God had created no Salt, the Art of Alchemy had not been. Salt is Coprose, and coprose is Salt; all lesser and greater minerals truly are nothing else but salt; nothing is more fluxible than salt; nothing more piercing than salt, and his nature; nothing cleaner, purer, more spiritual, and more subtil, than salt and his nature. Nothing stronger than salt and his nature; nothing more incombustible than salt and his nature; nothing more volatile than salt and his nature; nothing sweeter than salt and his nature; nothing sowrer than salt and his nature."

These two passages do seem to be repugnant, saying sweet and sowre, which is understood, Sowre before the preparation and Sweet after.

And following they say: "Nothing is nearer to the fire than Salt and his nature. Nothing more lasting

45

and fit to preserve things from putrifaction, than Salt and his Nature."

Thus seeing the salt even so as he is without other preparations, is of such virtue that it preserveth things from putrifaction, as is seen by experience; what will it do, when from it the elements shall be separated and reduced into a fifth Essence? I think with myself that it shall be that, which our Philosopher understandeth of the Secret Spirit.

Now let us follow. The Salt is the life of all Things: I cannot fail when any brave place of importance commeth to my hand, but I must speak my opinion agreeing the Philosophers together.

This Philosopher saith, "Salt is the life of all things." And Morienus saith, "But this Stone is not a Vulgar Stone, because it is more precious, without which nature worketh nothing at any time, whose name is One."

By the which I say that seeing salt is the life of all things, it is necessary to say and affirm with Morien, that without it nature worketh nothing at any time.

And Raymund in the Chapter before alledged, speaking of Salt under the name of nature, saith, Seeing this is, because nothing can be engendred without it, & etc. And I may bring hither many other Philosophers for confirmation of this wonderful sentence, the which I will omit, for to follow the rest.

"Salt is nothing else but a pure dry water; nothing more transparent, nothing more shining, nothing more

lightsome than Salt and his nature." If I would tell
my opinion upon all these sentences worthy to be
written in Letters of gold, I should enlarge myself
too much. But this I cannot let pass with silence,
for confirmation of so many excellent sentences
above spoken in the 2nd. Chapter, concerning the
light. And here is seen this Author to confirm the
same, saying: "Nothing is more transparent, & etc."

Now let us follow, "Nothing is nearer unto metals
than Salt and his nature." How is it possible to be
silent with this sentence worthy to be graven in
plates of Gold, and not written in Paper? O how open
a field is here to discourse! But let us follow,
"Nothing more excellent, created by nature, than
Salt and his nature. Nothing more simple than Salt
and his nature. Nothing more stinking than Salt and
his nature. Nothing more odoriferous than Salt and
his nature."

Seeing those two passages do appear to be
disagreeing, it behoveth to understand them as these
others above, if sweet and sowre; to wit, before and
after the preparation.

"Nothing better in nature created by God than Salt
and his nature. Nothing is in the world that
containeth so divers colours in it, as Salt and his
Nature. Nothing heavier and weightier than Salt and
his nature. Salt is of a nature, animal, vegetable
and Mineral and hath in his nature the actives and
passives." And here is verified the speech of
Aristotle, saying, "It is a Stone and no Stone, and
it is mineral, animal and vegetable, which is found
in every place, in every time, and beside every man,
& etc. Our Oil, Our Water, Our Sulphur, Our Mercury,

is no other thing in his virtue than Salt. There are three Stones of White Things, which three are found in Salt.

Salt is a Virtue mixt with all the Elements. There is nothing that so strongly containeth in it the four Elements as Salt."

I will say nothing upon this place, because who will study well all this which is above said, shall find to be here the last conclusion and Key of all.

Therefore let the mockers of Alchemy hold their peace, seeing without true dissolution they can do nothing, and true dissolution can they not have, without they reduce the Things dissolulable into the nature of Salt, and make them resolved that they may the sooner be resolved."

And, to put an end to this Chapter, I will with the help of all the Philosophers conclude, who say: "Therefore who knoweth Salt & etc., his dissolution he knoweth the Secret of the Ancient Wise men. Therefore set thy mind upon Salt. Think not upon other things. For in it only is hid the Science and the Chief Mystery, and the Greatest Secret of all the Ancient Philosophers.

The Fifth Chapter

The body now being reduced into the first Matter, and made fit, and disposed for the separation. Albeit many and divers philosophers have at large handled it, never the less I will shew one way clear enough, written by an uncertain Author, and Ancient, very pretty for the purpose of that Secret Spirit, with some addition or gloss of other philosophers for better declaration of the Mastery.

Now let us return to the Text, which telleth, "How that Spirit corporally appeareth, that it may be found the sooner and known & etc."

For declaration then of this second part, we will speak with the above named Author in this manner that followeth: "Take the blessed Stone which is no stone, nor hath the nature of Stone, and separate the Elements. And note that the philosophers calleth Stone all that from which the Elements may be separated by Art: For by conjunction of them in the work of Alchemy is raised a Substance like unto a Stone.

And he calleth it blessed, because beyond the four Elements there resteth a fifth Essence, called the Spirit because the Spirit cannot be seen by us, nor felt, without a body assumed in some Element & etc. This Spirit for the nobleness of his nature, taketh a body in a nobler and superiour Sphere, to wit, of the Elements; namely of the fiery Sphere of the noble fire, but yet his Spiritual nature remaining; therefore neither is it fire, nor hath it the nature of fire, so much as is of itself."

And again: "Because that body of the Spirit is fiery, for his subtility and purity, and this cannot be seen by us; therefore with fit instruments, by means of the workinans industry thickning its subtil substance, it is converted in form of water and floweth & etc. Therefore separate the said Spirit, and conjoin it with the Elements.

But the operation in the conjunction is two—fold; to wit, one that the Elixir may be made to congeal Quick-silver, another that the Elixir may be made for to preserve the life of man, and to throw away all superfluity of bad humours, and to eschew all corrupt ion of the body: Therefore if thou wilt make the Philosophers Stone to congeal Quick-silver, do in this manner.

Separate first the Spirit, and the soonest that thou canst, because thou shalt never separate him so warily, but that he will retain some part of the former substance of phlegme.

This Spirit once separated is called the burning water; whose sign is because a cloth dipt is altogether burned.

So have you one Element made spiritual, with the Spirit of the Quintessence."

And so the first apparition of that Secret Spirit becometh manifest in form of Water. The which is that water whereof the philosophers say: "The secret of the Art is to know the celestial water, divine, and glorious, & etc."

It followeth: "And so behoveth the other three Elements, to be made Spiritual with the said Spirit, retaining the corporeal virtue, in this manner.

Separate the whole superfluous phlegme from the said stone, until the oil comes to fume out, and nothing at all remain of the phlegme, and it shall be turned like unto pitch.

And then mix the first burning water rectified with this substance made like pitch, well stirred till it be corporated.

And then again distil twice or thrice, and then it is called mans blood rectified, and of this saith the Philosopher, the Wind hath carried him in his belly. And so have you two Elements exalted in the Virtue of the fifth Essence, to wit, Water and Air."

And this is the second apparition of that secret Spirit in the form of Air; of which another philosopher saith, "This is truly human blood, the true Celandine, in which the secrets of nature are hid & etc."

It followeth, thereafter take the foresaid substance, which remaineth like pitch, and separate all the superfluous oil by a glass Alembick, until that as oil remain. And then it will be a black dry substance, which powder well, and grind well with human blood rectified, and let it so stand for the space of three hours.

Thereafter distil, and then it is called the fiery Water; and do in the same manner the second time, and then it is called the fiery water rectified. And so have you three Elements in the virtue of the Quintessence, to wit, Air, Fire, and Water."
And so appeareth the third apparition of that secret Spirit in form of fire. But because this philosopher maketh no mention of the separation of one Element from another, and this I think is, because the separation of the Airy Element is not necessary to desire to follow the whole work. But who will separete it, to use it for the Virtue described by the Author of the Secret Spirit, many philosophers have told the manner.

But if you read the manner of the separation of the four Elements of Celandine, described by many Authors, and chiefly by Philip nestadius (Paracelsus?) in his Heaven of Philosophers, there you shall find the manner of the said separation, therefore I will not enlarge myself otherwise to write it.

And more I will say, that the Philosophers who hath written the separation of the Elements of Celandine, is that same of the Secret Spirit, who would serve himself under the name of Celandine, that is, the gift of the Heaven; and that this true, you shall find described the virtues of these Elements word by word, as those of the Secret Spirit in the fifth apparition outward. And the like did another under the name of human blood.

And as it is above spoken at full, it is no matter of the names otherwise, because all do understand one only Thing.

Now followeth here a most pretty glosse, worthy to be noted upon this passage of 3 HOURS. The which will give great light to the children of the truth, and it saith thus: "And in that space is melted all the White Volatile Salt, which is in the black earth with the foresaid water, and the water becommeth more sharp and burning; which whiles it is distilled, it carreath with it all this salt volatile and spiritual and flieth out in the stilling.

"The which salt is called fire and therefore this water is called fiery; of the which salt the names are these, the Salt of the Yolk of eggs, the Star Diana, the Morning Star, the flying Eagle the Secret of Nature, and infinite other names. Therefore Mercury is sublimed and made Salt; and so when you hear in the books of Philosophers anything of these names, know that it is no other Thing but the honoured Salt, and in it there are more than 50 names and so oft rectified until that it destroy all things by burning." Followeth: "After take the said black substance, and calcine it in a furnace of reverberation, until it become like lime, and with this lime mix the fiery water rectified, and distil it, and then it will be a water of life rectified.

And so have you the four Elements rectified, and purified with the Fifth Essence, and with the Spirit of the Fifth Essence, and this is the water of life, which is sought in the Work."

And here endeth the separation of the four Elements, with the fourth apparition of the secret Spirit in an earthly form; as more clearly appeareth in this gloss which followeth upon that place, (until it become like lime).

Which saith: "And this shall be when all superfluities and foul himidities shall fly out, and be separated by the flame of fire, and no otherwise; and the lime shall not be white, but black, rusty. And this is the true earth of the Philosophers, which is called the Secret of the Stone; in this lurketh the hid gold, and this hid gold cleansed from his earthliness and filth, I have touched with mine own fingers and seen with mine own eyes. For this earth excelleth all other earths of Alchemists; neither any doth hold in itself naturally the hid gold, but this alone. And therefore the medicine which is made by this is called one and sole & etc."

And so an end is given unto the Second Part of the apparition of the Secret Spirit in four corporeal forms.

The Sixth Chapter

In which shall be declared the Fifth Apparition of the Secret Spirit in a Glorified Body.

Followeth yet the same Author, and he saith: "And this water fixeth all Spirits, and maketh them enter; for this water hath her superiour and spiritual strength that is not fixed, and hath her inferiour and coporal fixed, and yet is not fixed, but hath power to fix." And this is it that the Philosopher saith, "That which is above is like that which is beneath, for the working of the miracles of one thing; that is it behoveth that this fifth essence, that is the Spirit, have or retain her spiritual power, and have all the corporal power of the four elements, if miracles should be done thereby; because if it have such power, many miracles are done upon the works of Alchemy." Also the philosopher saith, it ascendeth from the earth unto heaven, that is, the four elements have ascended from the earth unto heaven; that is to say in the spirit of the stone.

Thereafter saith the philosopher; "And again it hath descended into the earth; that is to say; these four elements have ascended into heaven, and again descended into earth; so that they be fixed in virtue of the Spirit of the Fifth essence, and remain one Crystalline Stone. And it shall be Elixir retaining perfectly Mercury of the fugitive slave."

And so is manifest the fifth apparition of the secret Spirit, under the form of a Crystalline or Glorious Body.

But here is to be noted, that this Philosopher sheweth not the manner particularly, how this conjunction of the Spirit with the body ought to be done; but only metaphorically saith; "They have ascended into heaven and again descended into earth; so that they be fixed, & etc."

Wherefore if I would produce the manner written by divers Philosophers I should be too tedious.

It sufficeth unto me only to say, that Raymund in the threescore and second chapter of his Codicil doth declare at large, and endeth in the threescore and fourth chapter of the said place. And in his repertory sheweth very well and clearly.

And here is verified the speech of Hermes, who saith: "The earth is the Mother of all the elements, and out of the earth they proceed, and to the earth they are reduced.

And Raymund in the Third Chapter of the theorick of his Testament saith, "And by this end everything shall go to his own proper place, from whence it first came & etc."

And here endeth the Third and last part of the secret Spirit, where he saith, "In the fifth and last essence it appeareth in a glorified body." And it is the desired end and true intention of the philosopher, when he saith, "But there is one Thing, which removeth all corruptions."

Now this Author (as I have above said) sheweth not the manner of the Composition (conjunction) of the elements, but under figures; and the like doth the Author of the Secret Spirit, but under other figures and another manner of speech, according to the letter; but as for the sense, it is the same.

And therefore to give the matter to the diligent to study, and to record one philosopher with another, I think good to set down here the metaphors of the said Secret Spirit, in which are these following: "Which remaineth fine in the Water, and carrieth with it the earth in the Air & etc." Therefore after he saith, "Containing fire burning in water, and earth clarified in Air & etc."

After: "The glorified spirit cannot appear but in a body agreeing to his kind & etc."

Furthermore: "Let not any man think, that the Spirits are divers; but howsoever it is called it is one and the selfsame spirit, who worketh all in all. This is the Spirit which in ascending the clearness of the heaven hath over shined, and in descending, the purity of the earth hath incorporated, & etc."

Albeit there be some others, yet they are not for this purpose; therefore they shall not be set down otherwise.

I say then, if you interpret well these sentences, with the others above alledged in the Mastnie of the separation and conjunction of the elements, you shall find them meet together, and agree very well.

Moreover, it is to be noted, that the foresaid author of the separation of the elements, maketh distinction of the aforesaid Elixir, for to congeal

Mercury, from the other for medicine to mans body; which the Author of the secret Spirit doth not. Also he saith, that it is a medicine fitted not only for both, but also hath many other virtues as in him you may see; and the like many other Philosophers do affirm.

Nevertheless, for to satisfie every man, as also because there are very many fine sentences, very necessary, appertaining to the foresaid Elixir to congeal Mercury, yet another manner shall be shewed for the health of man, as this Philosopher would, the which is this that followeth.

The Seventh Chapter.

Wherein is shewed the Manner to make the Elixir, or Medicine to conserve the life of Man.

"But if thou wilt make the Water of life, to conserve the life of man, and to cure all diseases, proceed thus: Make a burning water very well rectified. But make not of it mans blood, for if it were human blood, it would lose his force attractive of the virtues of herbs, by reason of his too much unctuousness; and would defile all the taste; and so would be unfit to be received by mans nature.

Also make not of it the fire water, because then it would be of so great force and sharpness, that it would destroy all by burning and so it should be dangerous to be taken and received.

But because the perfection of every stone, consisteth in the virtue of his earth; because it is called the nurse, or leaven unto it; witnessing the Philosopher and consenting, who saith:

THE NURSE OF IT IS THE EARTH: Without which leaven, the Spirit of the Stone can no ways be retained or detained perfectly, or possess the accomplishment of his virtue: Therefore; give unto this water the virtue of her earth, and then it shall obtain completely and intirely its own Virtue."

And this is what the Philosopher saith, "His virtue is whole if it be turned into earth: And then it is called the Water of Life, but if you distil it from that earth, it shall be the water of Life, rectified and perfected.

Know therefore, that in this Stone the earth is small and of great virtue. And care you not if there be little of the earth; because as a little leaven leaveneth the greatest quantity of the paste; so this little of earth which this Stone containeth, sufficeth to fulfill the nourishment of the whole Stone. Therefore seek not a strange earth, as some do, who take the Tartar of wine and say that it is the earth of the stone, some the dregs of Wine calcined, others the ashes of the Vine: And these do err, so the blind leadeth the blind, and both fall in the pit, believing to make the Water of life, and make the water of death. Because the earth must not be strange, witnessing the Philosopher GEBER, saying, one Stone, and one Medicine, to which no strange Thing is added, but all superfluities removed. So it is in this Water of life, NO strange Thing is added, but all superfluous Things are removed. Therefore this is the blessed Stone rectified.

Or the foresaid Water if it be thrice sublimed through his dregs, that when a drop thereof is put into a spoon, and kindled with a candle, it is all burned, so that nothing remain in the spoon; and then, it is rightly rectified, and this may be perceived in the preparation thereof, for this end that it may be profitable to cure diseases, and to conserve the life of man. And because the earth is necessary before, that the rectified water should be distilled from it, then it is necessary that the oil be drawn out, and separated from the earth. Know that the foresaid earth is all burned, and stinking

as burnt Things. And unless the foresaid earth will
be washed with the water of the phlegm, so that it
lose altogether his stink, the rectified water would
retain something of the stink, which must pass
through the earth, and be distilled from it. And
therefore before you make the Water of Life now
rectified pass through the earth, first wash the
earth well with the water of the Phlegme; so that
they lose well the stink of his burning.

Which done, from the said earth, that is to say,
through the said earth make the water of earth
rectified pass. And this you shall do at least 7
times, and then shall you have the Water of Life,
medicinal as I have above said. And note that scarce
can you have a pint of the foresaid water of life
rectified, out of thirty pints of wine.

Let it pass through the earth many times, and it
will be more effectual and his virtue will be
multiplied and increase, because the oftner it is
distilled through his earth, so much the more
effectual and powerful shall be the said water.

Item. Note that in the said water are dissolved the
leaves of most thin gold, and so of gold is made
Aqua Potables, and it is wonderful to conserve the
life of man, and to take away all diseases. And
which is more, it maketh old men young again.

Therefore regard warily the foresaid water. The
Water of Life above written, is sometimes made to be
Elixir or Medicine, to congeale Mercury. Sometimes a
part to be medicine for to conserve Mans Life: The
virtues which are communicated to everyone, we will
briefly set down in this Chapter.

Know therefore that the Water of Life, which is made to be Elixir to congeal Mercury, not only congealeth Mercury, but also blancheth Venus, and dissolveth Spirits, and Calcineth bodies.

But where it is made to be medicine for the conserving the life of man, you have his virtues and praises in other books; For it breaketh the impostumes, and cureth wounds from rottonness, & etc.

The simple water of life is drawn out of wine, and is called the soul of wine, and is called the Soul of Wine, whose glory inestimable, is the Mother and Lady of all simple Medicines and Compounds, whose effects are wonderful, and especially against all causes and passions of mans body.

There are many ways to rectifie it, but in the above written Chapter I have put the best inventions. When the foresaid Water of Life is distilled at least four times, there is no means to distil it from its earth, as I have set down in the Chapter above written, But it sufficeth that it be distilled as commonly it useth to be distilled, that the health of man's body may be conserved, and lost health restored. This water is so strong, and of so great virtue, and of the greatest natural heat, that by itself and without cominixtion, it cannot be drunken without hurt.

Item. If the eyes be weakened through a web, or for want of spirit, let there be put in wine the leaves of eye-bright, Rue and Vervane, of each one handful, of Celandine a little, and all being bruised amongst the hands let them be put in wine, and stand there over night, and thereafter the foresaid water of life be distilled.

Item. If the herb Balm-ment be put in wine, and afterwards the water be distilled, then one spoonful be taken with a fasting stomach, it maketh a man well remembering things passed, and retaining things heard.

Item. If Sage and Mint be put in wine and thereafter the water of life be distilled; the water drunk, killeth wonderfully all kinds of worms.

Item. The water of life made with Terpentine cureth the quartane Ague, if it be taken before the fit, and make water also with it.

Item. Note that whatsoever odoriferous Powders, or whatsoever green or dry herbs shall be distilled in the foresaid water, it shall retain the smell, and shall be powerful. And the user of such water shall feel the power and virtue of these spices, and if guests chance to come, wine being mixed with the foresaid water, incontinent shall retain the taste and smell of the Spices and herbs put into it, and so it shall appear to be Clove Wine or Sage Wine, & etc. And so every discreet, wise and understanding man may seek out the virtues of the water.

And note that all which are written, to wit Medicines, you understand the WATER OF LIFE, which is called the burning Water, and is the greatest subtility of wine or Spirit of the Soul. And the second water which is extracted or sublimed from wine by the same manner, is the Element of Air and burneth not, but some call it the water of Phlegme, because it is of a cold nature, wherefore I will describe some Things after this sort. If you will make hairs yellow, make ashes of Ivy-wood, and make a lee of the foresaid second water. Thereafter often wash thy head with the aforesaid Lee, and know that in two months the effect followeth, and it will kill all kind of worms that is in the hairs.

Item. The washing cleanseth the face, for if the face be washt therewith, the rose got (or sauce phlegme) is taken away. And if oil be made of the inward Kernals of Pine Apples mixed with the foresaid water, it shall heal and cure and cure it quickly; chiefly if the said oil be applied in hot-milk and that this be done with the swimming above.

The Eighth Chapter.

Where are to be handled the divers Workers in this Science.

Seeing that hitherto by the grace of the Lord God an end is put unto the Exposition of the Secret Spirit; and it is shewed by many sentences of divers wise Philosophers, the great difficulty and deepest depth of so high a secret, not only in knowledge of the matter, but also in preparation thereof: Wherefore the Philosophers say that it is very difficult, and they do speak in this manner.

The Philosophers have hid the preparation of the Stone, because it is the Key of the Art, and difficultness of Things.

Some others say: "The Working and government thereof cannot be known but by the gift of God or instruction of a Master who should teach it."

The same saith our Philosopher of the Secret Spirit.

Therefore that would I know, what we should think and say, of so great a multitude of men (which otherwise I know not how to name) that when they have seen a simple and sophisticate recipt, say and affirm with an oath, that they can make the precious and most beloved stone of the Philosophers; the which they have purchased with so long studies, troubles, tears and sweats. Which is altogether against the use and order of all the Philosophers, as Villa Nova saith, "The Alchemists of latter time, are for the most part mockers, and whiles by sophistications they seek rather to seem wise then to be, they deceive the yeilders to them, but the ancients not profiting according to their own covetousness, have wrapped up this Art in riddles, shewing rather their own ignorance then science, & etc."

I say then what should we say of these? Truly it cannot be otherwise answered but with the conference of JANUS LACINIUS, and Bonus Ferrariensis in his precious new Pearl, to exclaim and say "And no wonder because it falleth out of his desperate age, that men of every sort, and some of the most ignorant, dare search the hid causes of the Art and Science of this most happy and most high Philosophy, thinking to wrest and steal that blessed stone out of paper tricks, and deceits of some idiots: for they are smiths and weavers, carpenters, and such kind of men, desiring to be enriched without labour." An answer certainly to the purpose, worthy of such kind of people.

But moreover, what shall we say of an infinite number of lettered and learned men? Of whom I know enough, who are searchers of this Science, and nevertheless understand not the most obscure books of the Philosophers, to be written under Metaphors, but as the letter standeth; and consider not, or else will not consider what the philosophers say:

"We have not written our Books but unto our children, and our children are they who understand our sayings."

And Plato saith, Who knoweth our purpose, and our intention is now a Philosopher, and is inriched; and who knoweth not our sayings, he is in the suares of nature, & etc.

And Geber: For where soever we have spoken plainly, there we have said nothing, but where under riddles and figures we have put something, there have we hid the truth.

And Arnuldus: But the foolish understanding the sayings of the philosophers according to the letter, are become blind, and have found nought but a lie; and then they say the Science is false, because we have tried, and find nothing, and then they are like desperate men, and do despise the books and the Science, and therefore the Science despiseth them, for our Science of the hid things of nature, hath no enemy but the ignorant.

Therefore this divine Science is not purchased by being lettered and learned only, seeing it is the secret of God, as all men do affirm, for the which it is written; "Because all Wisdom is from the Lord God, and therefore sometimes these Things are given to the simple which the most studious cannot know."

Now let us leave this. I could in particular tell some manner of working of many, which I have seen in my Journies, of divers workers, which I will leave, that I be not too long and tedious.

But I will tell Two Ridiculous fables; which I have seen in this noble City of London, where I was present myself, of two of my best friends, searchers of this Science.

The one of which having divers ways tried fortune, and being one day by chance in a very ancient Palace, where he saw a glass window, in the which was painted the history or fable of JASON, When he, went to COLCHOS, to purchase the Golden Fleece.

Where reading something written, a fantasie entered his head, so that he would not understand that the Philosophers Stone was made of other then of glass, alleging a sentence of a Philosopher saying, In salt and glass is all the Secret.

And again he said that Alsidius speaketh, "Break the glass and extract the Stone, and put it in a glass vessel, or bolts-head, and extract the oil from it, and you shall find this which the philosophers delivered unto us, in this glass is the Quicksilver which overcometh the fire, and is not overcome by it."

And Raymund confirmeth the like in the 86th. Chapter of the Theorick of his Testament, where he saith, "Draw the quicksilver out of the caves of glass & etc." But what more?

He alledgeth two passages of the Revealation of S. John the Evangelist, Chapter 21. Where he saith: "And the building of the wall thereof was of Jasper Stone, but the City itself, pure gold, like to pure glass, and furthermore after, And the street of the City pure gold, like transparent glass."

Where I remaining a little wondering, at this his fantastick fantasie, asked him what affinitie and friendship, and what to do had glass with metal? He answered me that I understood not, and that it was understood the glass made of Metals, alledging the speech of the philosophers saying: "That the glass of Metal changeth every Metal better, & etc."

And John Bracesous understood the same in his dialogue of demo-gorgon: And Geber, when it is said, "That glass made of iron is the Philosophers Stone."

And so likewise may be made glass of gold and silver, and of all other metals.

Wherefore leaving him with this his Chimera: I will speak of the opinion of the other no less fabulous than this.

I say that this my other friend said and affirmed, that he had the Knowledge of the true Lunaria, so much mentioned by the Philosophers, and that in it did consist all the secret of this Art. Out of the which (as he said by a Philosophical way) he did take the Juice, and of which he made a Salt, which was green, saying that this was the true Salt which the philosophers understood, alleging a sentence of Hermes, that saith, all Salts of what kind soever are contrary to our Art, except the salt of our Lunaria.

Of what salt he said, by divers operations he took the Mercury, the which was the Mercury Vegetable; of which afterward he separated, not only the four elements, but also he took a Water, which he called the Spirit of the Stone or fifth essence, alledging an infinite number of sentences of divers philosophers for his purpose.

And chiefly Raymund Lully, and principally, in his apertory, where he saith, "Take of the best Juice of Lunary which thou canst find, & etc. And the Rosary of the Philosophers, where it saith, "The Juice of Lunaria, the water of life, the fifth essence, the burning wine, the Mercury Vegetable; are all one, the Juice of Lunary is made of our wine, which is known to few of our Children. And with it, by the means of it, is made our potable Gold; and without it no ways."

And more he said, that after he had taken his Mercury out of the same dregs or earth, he could take as much of the same Mercury as he pleased, without end, the same earth remaining never the less ever in his proper weight and quantity as at the first: Which appeared wonderful unto me.

And I asking reason thereof, he answered me with Vencentius in his natural, "The Light hath the property of the Fountain, the cause of Multiplication." And more he said, that this his earth was like a well of such water, as never could be dryed, and it was the body understood by John Augustine Pantheus, in his Voarchadumia, where he saith, "That the Vegetable body is full of Juice & etc."

And moreover he said, that this was that true Salamander, that was engendered, and nourished in the fire, alleging many authorities of Philosophers amongst which he made use of a book entitled, "The Water of Life Perpetual," which said, it is fire of fire, and is engendered of fire, and is nourished in the fire, and it is the daughter of the fire, &etc."

And that more he said, that it was also that thing, and the Spirit of the World which Heriri255 Cornelius Agrippa speaketh of, in the fourteeneth chap. of the First Book of his Philosophy, where he saith: "But it is more infused into those Things which have taken most of this Spirit.

For it is taken by the beams of the Stars, according as things render themselves conform unto them. Therefore by the Spirit every hid property is propagated in herbs, stones, and metals, and beasts, by the Sun, by the Moon, by the Planets, and by the Stars higher than the planets, yea, this Spirit may be more profitable unto us, if any man know how to separate him well from othe elements, or at least use those things which abound most in this Spirit."

So that he made me remain so confused, that I knew not what to answer. Whereby I am disposed to stay no more with these melanchol ick and fantastick humours, that I make not myself fall into some Heresies to no purpose.

For the which I will exhort the true searchers of this noble science, that they suffer not themselves to be fooled with vain opinions, nor to set a work in the day that which they dreamed in the night, as these two my foresaid friends have done.

But to be constant and follow the documents of the foresaid good philosophers, and so shall be made an end of this my short discourse, which is dedicated and presented to the Children of the Truth, who delight in a solitary life.
Now my dear and rude book, thou hast endeavoured to set forth all thy will in speaking and declaring by

a method, and continued order. Gathering together so many fine sentences, described by so many worthy and wise philosophers, and scattered not only in many chapters, but in divers books, against their precept. Who do command that this noble Science should be written obscurely and not with a clear and continuated stile, to the end it be not usurped by the ignorant & unworthy people.

But seeing that so it hath pleased then to do: At least flie from the multitude of men, and learn the solitary life: And converse only with those noble and solitary spirits, to which thou art dedicated; because in the solitary life is found this most noble secret spirit; secret it is called because also it truly shunneth the conversation of the Vulgar, and goeth to hide itself in solitary and secret places.

And moreover, because that in the solitary life is learned to know God; In a solitary life, is learned to love God, in a solitary life, (I say) is learned to give Glory and praise to the most high, and most glorious Creator of all, to whom be praises through infinite ages of Ages.

Finis

TRIFERTES SAGANI

OR

IMMORTAL DISSOLVENT

BEING

A Brief but Candid Discourse of the Matter and Manner of Preparing the Liquor Alkahest of Helmont, the great Hilech of Paracelsus, the Sal Circulatum Minus of Ludovicus de Comitibus: or our Fiery Spirit of the Four Elements.

TOGETHER

With its Use in Preparing Magisteries, Arcana's Quintessences, and other secret Medicines of the Adepts from the Animal, Vegitable or Mineral Kingdoms.

BY: Cleidophorus Mystagogus

2.   Maccab. ch.1.v. 19. to the 23: The Fire of the Altar turned into thick water.

2.   Esdra. ch. 14.v. 39. And behold, he reached me a full Cup, which was full as it were with water, but the Colour was like Fire.

AN EPISTLE TO THE READER

Various have been the Opinions (Courteous Reader) concerning the Basis and Foundation of this general Dissolvent, commonly known by the name of the Liquor Alkahest; some imagining it to be Mercury Prepared; others, that 'tis in Urine, Blood and the like; which has been the reason of so many and difficult labours made use of, and all in vain; for that the Liquor remains at this day as great a secret in the world as ever, and 'tis like so to continue while Chemical Authors deliver the subject in such Tropes and Metaphors, which horrible and inextricable labyrinth the young Tyro's are so entangled in, that it must be by more than an ordinary Providence, that he can be disentangled and set free.

On this account it may be properly said, that these Chemical Writers had as good to have been silent; nay, 'twou1d have been better for then so many would not have engaged in a search, where so little likelihood of obtaining was seen whereby a great deal of Precious Time and Money might have been saved, and that Perplexity of mind which follows vain Chemical Processes might have been Prevented; for this Reason it is but just in Authors in all their discourses, tending to the instruction of others, to direct to that Subject which is the true Object of that discourse; and tho' I must acknowledge that 'tis not fit to be delivered or disclosed so plainly, as that every Hog that may come to the Honey Pot, yet I say that it may be Clothed with such a decent habit, as to prevent it Abstruse enough, yet by this a certain and harmonious Concord to be seen; as for example, the

Object of this discourse is the Alkahest. Man and all creatures have it, for there is no being in Nature, that is rightly and genuinely dissolved, but what may properly and truly be said to be done by this Liquor, but particularly in man after a more evident manner in all Chylifications whatever, but in this Act Man sucks the Quintessence of all things so dissolved for his own Nutrition and being Transmuted into Human Species, the Recrements are cast off by the common Emunctories, far more grosse and Imperfect, than the Species themselves were in their first Reception, consequently they are of no fit object to ground the discourse of so pure and Immortal a dissolvent on; nor indeed Man himself, tho' we grant that it is plentifully in him, but 'tis that Salt or Life in him which Concentrates all other Salts in his own Essence as a Catholick Fountain for all the Rivulets to be supplied from. So that there is no taking of it from him but by a violent breaking of the Glass and a Transplanting of the Fountain back again to that Inexhaustible Ocean from whence it first received its being. Therefore, man cannot be the Object of any discourse appertaining to demonstrate the Subject of this Dissolvent; for the pure in him, as already said, cannot be obtained without Death, which is abominable even to think on; but if it were obtained, it would not answer the end; for what is sufficient to dissolve in the Vegetable Kingdom is too weak for the Animal; and that which is strong enough in the Animal is too weak for the Minerals. Therefore, seek it in that and from that, which is the Fountain, that supplyes all Creatures and beings with it; for had it not a Source, Nature would soon cease, for as she exhausteth by the Acts of Motion and Agitation of Parts in Generation, so is she on the other hand Immediately Supplyed, not only in the Great World, but also in every Individual Part where Life is: Receiving the same through the Air, as the true Vehicle, consequently this Catholick Fountain is the Right object to ground this discourse on, as the true Subject of the said dissolving Liquor; this is a standing Truth, 'twas Truth in the beginning, and will remain so to the end of Time. The Reason and Philosophy of it I can by Mechanical Demonstration make clear to a person worthy of such inspection: therefore by all Clouds, dark vails and Metaphors, I genuinely declare that the matter of this dissolvent is one and the same in essence with that matter from whence all the wise Antients obtained the Universal Medicine one being the Work of Art, the other Nature. One gentle, the other violent: so that by the difference of Operation they are brought to different effects.

Now therefore the Subject of this discourse being the Liquor Alkahest, the object must be the Universal spirit; for 'tis from this grand fountain of nature that our chaos doth proceed; therefore thrice happy is he, that knows those magnets that attract and make a species of this general Genus: for be assured that there is something more than elements in all created beings, even an incorruptible and quintessential spirit, which is the very life of elements themselves; which being taken appears in mist, vapor or water, even that out of which the Ancients say all things were generated, However, the right knowledge of this matter is sufficiently abtruse, and the operations thereof yet more abtruse, for I with many others know by experience, that the matter may be known and many

doe know it, yet are wholly to seek in the matter
and modus of operating thereon; and whatever some
foppish and conceited ones may think, viz. that if
they had the knowledge of the true matter all
difficulty is over; this may prove a grand mistake,
for I have been intimately acquainted with some,
that have had a true knowledge of the matter, and
have wrought thereon; yet to the day of their death,
have been to seek of the magistery.

These difficulties have been those sharp stones that
have hindered my legal progression in twenty years
travel towards the mount helicon of art; so that my
labor of body has been excessive, and that of the
mind much more. I have passed through the drought
and scorching heat of the day, and also through the
cold and chilling frosts of the night through a
multitude of wants and difficulties even often to
the hazard of life itself. How easy soever such, as
aforesaid, may think it: I have had no other door to
come in at, but hard labour and great expense, for
coals and glasses have been my interpreters, and
shall be so to every true son of art to the end of
time. So that for conscience-sake, I'll write the
truth as well knowing that there are too many
sophisticated and false proceses in the world, which
will not bear the touch-stone of experience; but
vanishes like a reprobate metal upon the test.

But passing by all this, and much more that might be
said of the like nature; I shall now address myself
to the desireres of Wisdom, and let them know, that
I have, as in a Glass, showed them the true manner,
in which and by which this Dissolvent is to be

obtained. I shall now come to speak concerning its Use, and Utility when obtain'd which will abundantly reward the Possessor for all his costs and paines bestowed about it; for as it hath been deliver'd, that the subject of this menstruum is universal, so are the acts of it the same when prepared, which plainly demonstrates from what fountain it might flow.

This liquor, as an universal fire, dissolves and opens the textures of all beings, in the vegetable, animal and mineral kingdoms, into their nearest matter, which is saline, sulphureous, aqueous and potable, diffusive in any liquor, and so comes immediately to nature's relief, and by the specifick virtue manifested from power into act, diseases tho' never so deplorable, may be overcome and cut down, as grass or weeds with a scythe in the hand of a mower; especially by those of the mineral kingdom, which may be justly esteemed the physician's crown and philosophers diadem. This is the liquors virtue in general. In particular, as it universally acts without limitation on all subjects in the world, so in this action there is something remarkable to be observed in every subject; for it fixes volatile spirits, and volatizes fixed ones. It makes salts sulphureous, and sulphurs saline. Nay it macerates the gummosities of resinous and gummy things, which the ferment of our stomachs could never do; for it being distilled from amber and terpentine[2] leaves them in a salt of excellect virtue: from the latter I have observed it almost as sweet as honey, and a powerful specifick in the stone; therefore by the help of this liquor of fire a few medicines being

[2] See Urbigerus "Circul. Minus" -HWN

prepated will answer in deplorable cases all that the patient can hope for and the true physician expected to perform. For instance, terpentine, so dissolved, or the stone ludus. Infallibly cures the stone in the bladder; amber and hellebore hysteric fits, hypochondriac, melancholy and madness. Cinnamon Unicorn's horn and the liver of an eel, for the speedy delivery of women in childbirth; the sulphur of Venus is an Universal Nepenthes, without opium in all diseases: the lilly of antimony for dropsies and all agues; the magistery of gold for malignant Fevers Pestilential Palsies and plagues: as also the glorified sulphur of the metallus masculus, by Paracelsus called Vinum Vitae and Membrorum Essentia, which also cures consumption, fixt Mercury or the arcanum coralinum and horizontal gold in leprosies, gout, palsy, epilepsy, cancers, wolves, scorbite, Kings evil, all sorts of the venereal disease without salivation or detaining the patient from business; 'twould be too long to enumerate all those medicines prepared by this fire or liquor; therefore ket these suffice, and the reward that may accrew here from, because by this way of practice, the physitian may justly and conscientiously gain honor and riches; and the patients be freed from all those cruel barbarities, which are the Adoequate parts of common practice; as if the pain and terror of death were not enough, but there must be an additional cruelty, viz. of blistering, which to some may exceed the former.

Thus having given you a short Scheme of the Rise, Preparation and Use of this Liquor, I shall Conclude this Preface in Consideration of a Person rightly qualified for the Possession hereof: the first and most necessary Qualification is to be rightly Informed in Religious thing's so as to know God for themselves savingly, by passing through the Holy River of Regeneration, for to walk in the newness of the Spirit, which Divine gift of the Holy Ghost enables every true Christian to walk with that Circumspection, as to be acceptable in the sight of God, to such it is a sure Guide and safe Conductor in this World towards the desired Haven of rest. It is also to them a mouth and wisdom, and that by which their Tongues are bridled and the whole Man Sealed to the Day of his Redemption, giving an earnest opinion the Eternal Inheritance, and afterword's a dull Possession, when our Mortal shall put on Immortality; this being the Fountain, all others that are true, follow as Rivulets from it, and so give a right Qualification for the knowledge of nature and natural things as also a constancy of mind to work upon the one thing alone, and an industrious hand to effect the same. Here a blessing and success may be hoped for, and those incredible rewards for all such as wander in the circumference, and have never been admitted to the center of things; but to the vigilent hidden things, even those hid from the foundation of the World, shall be reveal'd: and that this may be the portion of every true Laboror in Art is the sincere desires of him, who wishes the General Prosperity of Mankind every way.

                              Cleidophorus Mystagogus

CHAPTER I

CONCERNING THE MISTAKE OF THOSE WHO HAVE SOUGHT THIS LIQUOR IN WRONG SUBJECTS AND BY WRONG WAYS

It is a Saying worthy of Observation, that the Industrious Hand makes Rich; so is it in all manner of Trades and Convers in the World; so is it in Art; but this Industry must be upon a tight Foundation, and, in the Chymical Art, from a Fore-knowledge of Adequate Causes; unto which it is impossible to attain without we are enlightened by that Wisdom, which comes from above, as a Ray from the Holy Heavens and Throne of the Divine Glory; for 'tis she, that must Conduct us in all our Labours to make them Acceptable to the Great God; well therefore might the wise Man esteem of her before Riches, and Prize Understanding above the Merchandise of Silver, Gold, and Precious Stones, because she is the true Conductor to the ways of Peace and Pleasantness; nay even to that Tree of Life, where Substance is to be Inherited: For that she opens the Door of Entrance to all Mysteries Divine and Natural; and consequently without her Men grope, as it were, in the dark, even as a Blind Man does at Noon-day; for Nature God's Hand-maid was Created by him, and Job says, that God by his Spirit has garnished the Heavens, his Hand has formed the crooked Serpent; and tho' there is a Spirit in Man, yet 'tis the Inspiration of the Almighty that gives Understanding, whence we may readily Conceive, that Human Reason is too short to Comprehend the Dignity of any true Mystery without the Aid of God's Spirit.

This great Defect is too evidently apparent from the deplorable Case of the Chymical Searchers, concerning the Subject Matter of this Discourse; seeing they know not where to ground or fix their Intensions in the choice of a proper Subject, but frames each to himself a different Basis, and so make an Innumerable Number of Errors concerning the same: This Imaginary Matter, which Phansy only has given Birth to, they defend with all the eagerness imaginable, concluding it to be the Genuine Offspring of Truth; when, alas 'tis but a Bastard Brat of their own wandering Imaginations and ungrounded Thoughts, as in the Conclusion proves too Evident: This is an Absurdity so great, so common, that amongst the many Pretenders, I have never met with more than three that have escaped it: How then can it possibly be expected, that such should ever Arrive at the wished for Haven of Rest, when Ignorant both of the way and means by which they must come thither; for the Door of Entrance must not only be known, but also the Key which opens the same, without which they may never expect Admittance into Nature's Treasury:

Therefore consequently must still remain in the horrible Mist of Errors; the most principal that have come Athwart me I shall here lay down and reckon up for Convincing of the Giddy Headed and Rash Searcher, but more Principally for the Edifying and Building up of a Son of Art.

I shall begin first with an Error, which is almost Universally received, viz. that Mercury Vulgar is the Foundation or Basis of this Liquor; this is an

Error that the Authors of some Expositors are guilty
of, which the Ignorant Searcher has not been aware
of, but hath gone to work as confidently on Mercury
for the obtaining of the Liquor Alkahest, as others
have done for the making the Mercury of the
Philosophers, by several and various Preparations,
as endeavouring to break its Body by Spittle, May
Dew, Vinegar, and such like soppish Proceedings;
also by Sublimation with Salts, and Distillation,
and other such like Operations, endeavouring to make
it run *per deliq*. to obtain an Airy and Universal
Nature and radical dissolution, even that they call
the Magnetic Salt, or Foliated Earth, and Mercurial
Chalybs; but all in vain; for that Mercury so
prepared is still all one with common Mercury: And
so likewise is that, prepared by Regulus of
Antimony, Silver, and *&c.* for Vulgar Mercury is
unripe Fruit fallen too soon from the Tree,
therefore it must return to its First Fountain or
Catholick Mercury to be dissolved its self;
consequently is not the Subject of this Liquor; for
the Philosophers Introduce Fire, not Water, into
Mercury, to make her Medicinal, both in the
Particular, and also in the General; by which 'tis
brought to be forever Irreducible to Mercury.

Another Error is in those, who seek for this
Dissolvent in Dew and Rain Water, not considering,
that this was designed only as Nourishment for the
Vegetable, having but such a Portion of the
Universal Fire in it, as might serve to dissolve the
Salt Nitre of the Earth, and then the Vegetable Seed
in order to a new Production; this Fire or dissolver
being far remiss to that of Animals, as that of
Animals is to Minerals, cannot be the Philosophers

Subject of this Dissolvent, for Life would be too short to extract it.

Another Error is that many allow the Matter to be Universal, but is drawn by certain Magical Magnets at select times in the Year; but this is a grand Error, for the Matter is to be found Plentifully at all Seasons of the Year, especially in such places, as are mostly enriched by Mineral Fumes, and the manner of its Attraction is rather for the Necessity of Human Life, than any Point in Art; so that the Artist must not be too curious in endeavouring to perform that which Nature hath already done to his Hand.

Another Error is in those that seek for this Dissolvent by attracting the Air with Alkalizated Salts, as Tartar, and &c. not considering, that all Alkazated Salts do only attract a Saline Aquosity, which by often Cohobations may be turned wholly into an Elementary Water, whereas the true Philosopher (as already said) does by his Magnets attract a Fire, nay a Fiery Spirit stronger than any Fire in the World: 'tis true Alkalizated Salts are noble Subjects, and deservedly claim Preheminency, being Contradistinct to all Acids, and therefore make a Dissolvent next to the great Liquor; but these can never be Volatized without the Universal Medium, or Philosophers Diploma, together with Essential Oyls and Vinous Spirits, and being so Volatilized, they become noble Spirits, yet do notwithstanding spend their Virtue in Dissolving Bodies, and Coagulate upon them into a Salt, retaining their Volatility;

so that consequently these are excluded from being the Subject of this Immortal Dissolvent.

Another Error is in seeking for the Matter of this Liquor in the Animal Kingdom, viz, in Man; and indeed a greater in those, who assume to teach others, by their Assertions, that it is there; but having already detected such Writers (in the Preface) and also clearly shewn, that from Man, the Subject Matter of this Liquor can never be attained, altho' I know that this my Assertion does much thwart the general received Opinion, that Urine is the Basis, and that Van Helmont, Philalethes, Starkey and C. have in their Writings asserted the same, so that I do Contradict the Testimony of these Worthies: Instance Helmont, where he speaks of the dissolution of the Stone Ludus, seems to Assert that it is performed by a Second drawn. from Urine: and Philalethes, in his Treatise extant, has grounded the basis of the Immortal Dissolvent on Urine and Blood; and George Starkey in his Treatise of this Liquor seems to ground the [3] thereof on Urine; nay, an Intimate Acquaintance of his did affirm the very Process to me, which he made use of, viz the Urine of sound Men, unfermented, which, as soon as it was made, was by Evaporation brought to a Consistence, in order to unite the two Salts, Volatile and Fixed, and so by Distillation and Cohobation till the whole was brought over, and then being digested and dephlegmed the Alkahest is prepared. Dr. Bacon was, as I have been told, much of this Opinion; but all these are short of Understanding the Truth of the Subject, or of the Authors before mentioned; for it is easy to be collected from Helmont, Philalethes

---

[3] A symbol may be missing here in the R.A.M.S. work. -PNW

and *&c.* that they never depended upon Human Urine as the Subject of this Immortal Dissolvent for then they would not have directed you to the Chaos of the Ancients; as the true Subject, describing it Figuratively, and Analogizing it with Man; because Man Subsists by and from the Universal Spirit, which is the true Subject of this Dissolvent which they for some secret Reasons would not be so Candid to deliver: the like has Alipili in his Book Intituled, Centrum Naturas Concentratum, which very Title shews, that it is not Man there meant, but the Universal Spirit that being the very Life and Centre of all Centers: Therefore who ever shall assert, that Man is the Basis from whence this Liquor is obtained, let him be respected of Envy or Ignorance; because there is no Subject to be drawn from Man, that will act on Minerals five hundred or a thousand Times, and retain the same Virtue, quantity and quality, as if it had not acted at all: Therefore I regard not such Fops or such others, that dote upon highly rectified Spirit of Urine, mixed with the true Spirit of Wine, until both Coagulate into a Salt, which is Distilled and Sublimed by the Addition of fresh Spirit of Wine, until they come over in Form of a Fiery Liquor: There are others also that dote on the strong Spirit of Urine united with the Spirit of Vinegar, and Distilled into a Neutral Spirit: but Experience the Mistress of all true Art shews that these are all greatly mistaken, and many others, too long here to Enumerate: Therefore shall pass them by, and only Insist on some few others that remain.

Those are also mistaken, that depend on Acid Spirits, as the Subject of this Liquor, as Nitre,

Vitriol, common Salt, Salt Gen, or the Mother
Liquors of any of these, or any other Salts growing
in or extracted from the Earth; for all of them,
none excepted, will by Distillation yield an Acid
Spirit, and our Liquor being no Acid, but
Contradistinct thereunto, these of Course are all to
be rejected, and ought so to be in the Use of the
Liquor when prepared: others, that think themselves
more prudent dote much on the Spirit of Verdigrease,
and more especially if it is First often dissolved
in Spirit of Vinegar, and made transparently Pure,
and then shot in Spirit of Wine and so Distilled,
they then put as great a Price or Value on it, as in
Reason can be set upon the Immortal Dissolventits
self; but this Menstruum being Published by Zwelfer,
and long before by Basil Valentine, whom I take to
be the right and true Author of it, and being easy
to be prepared, it follows, that the Liquor Alkahest
would be no uncommon or unknown Secret; but that
remaining still as the greatest of Secrets, plainly
demonstrates, that these are not the Subject, whence
that is obtained.

They are also misled, who depend on Mineral
Sulphers, or the Vitriols of Metals or that of
Venus, described by Polemanus; because there's none
of these, but what are sluggish in themselves, and
unactive Beings, and can't be radically opened and
separated from their Mercury's, without the Liquors
help, and then they become Passive Medicines not an
Active Menstruum, so of Course are to be excluded
from being the Subject Matter of this [4].

---

[4] A symbol may be missing here in the R.A.M.S. work. -PNW

Another great Mistake and grand Error is in those, that depend upon the Essential Oyls, as Wormwood, Mint, Time; or the Oyls of Gums, as Amber, Benjamin, Turpentine, and these being Chaos'd down and devoured by Corrosives, as Oyl of Vitriol, Aquafortis, and being again revived, then be coming (as they say) the Regenerated Spirit of Wine of the Philosophers; which being Distilled from Tartar, Sal Armoniac and Mercury, each distinctly, till their Bodies are brought over, they are then the Magi's three Universal Menstruums, viz. Minimum, Minus and Majus: But this mistake has proved too evidently false, to the great Expense and Disappointment of many worthy Persons in this Kingdom, and indeed no better can be expected from such Heterogeneous and unnatural Mixtures, as being farther Alienated from the Universal Spirit, than some others already Named, and consequently the more remote from being the Subject of this Liquor.

To be short, I do on an Experimental Ground Exclude Animals, Vegetables and Minerals in all and every particular *Classis* and part thereof from being the Subject of this Liquor; therefore shall omit any farther Discourse of this kind and come nearer to the Matter in Hand, which is to detect the Errors of those, who confound this Liquor with the Mercury of Philosophers, saying they are the same in the Subject Matter, Identity and Operation; 'tis true, the Mercury of the Philosophers is a natural Dissolvent, but it dissolves in the way of Generation, when as this Circulated Salt or Alkahest dissolves it by way of separation and destruction; so that they differ in Operation, as much as Love and Wrath; the one in Love preserved, the other in Wrath destroys the Life and Motion.

There are also other Ignorant Boasters, who confound them together, yet know neither the one nor the other, yet say, they are both the same in Composition and Digestion, but near the Birth of the Royal Babe, the Matter divides its self into two distinct Parts, the one a Body Permanent, the other a Menstruous Liquor or Blood, which being Distilled is the Alkahest, this shews their great Ignorance, for the same that is a Body is a Spirit, and the Blood is Homogeneous with both the Mercury of the Philosophers and Liquor Alkahest; for the Mercury can never be prepared without its Aid, as being one of the three Springs; neither can the Spirit of the Body Subsist without the Blood, as every true Philosopher does, with me, know, and that at this State there is no division to be admitted, without a Death to the whole Compound; for the whole Matter in the Production of each being diversely wrought on produces the different Effect; the one is a Mercury Homogeneous, the other a Ponderous Saline Liquor and in the Production of both there are superfluous Oyls separated, which tho' Medicinal, are not in the least Homogeneous to either; which clearly Evinces their Ignorance in the Process of Nature, which is to make Bodies Spirits, and Spirits Bodies again, and that this Menstruous Liquor or Blood is the Life which is sown in his own Womb of Mercury for the Exaltation of both; for there the Heterogeneous Faeces are cast off, and so 'tis qualified and united with the Spirit in order to Redeem the Body; and so is a principal Ingredient of the Stone, when as the Alkahest is not: It would be too long to enumerate the vain and false Conceptions of Men concerning this Immortal Liquor, and seeing these distinctions do better become that Chapter, where the difference is shews between the Liquor Alkahest and Mercury of Philosophers, I shall omit speaking of it any farther in this, and come to shew the Subject of the Dissolvent in the next Chapter, and so Conclude this.

CHAPTER II

OF THE TRUE SUBJECT MATTER OF THIS DISSOLVENT

In the former Chapter, I have laid down the Mistakes concerning the Matter of this Immortal Dissolvent, which Helmont Describes in the Word *Latex*, which properly Imports an hidden Source or Fountain, so hidden indeed, that he himself says, when this was found Religion stood amazed, and well may the Religious Man be so indeed, when their Descriptions are so Occult; for from the Word Latex, which in Vulgar Reception signified Liquors, which may be properly conceived to be Aqueous and Spiritual, he presently comes to tell you, that the Masterpiece at which Art is Levelled, is to find out a Body; which may play with us in such a Symphony or consenting Harmony, by Reason of its exquisite Purity, that no Corruptive Principle can find in it any Heterogeneities by which to work in it a Dissipation of Parts; here he immediately calls it a Body; hence we must for certain Conclude that this Source or Fountain, tho' liquid, does contain a Body in it, or else it would have been vain in him to have directed us to such a Body for the Object, and that so Circumspectly and diligently, as to find it by hard Labour and Industry, saying you must be careful, or sedulously Industrious about finding out such a Body, which by Examen and Proof is very difficult to be found, because the Words import, that there is no such Body in all Natural Beings, that does Answer what this great Philosopher describes of his, therefore we must Conclude, that these Words also import Art; for that Industry is also recommended, which is as much as if he had said you must seek for

the hidden Source or Fountain of Nature, and
Universal Spirit, which Art must form into a Body;
but this Son of Wisdom doubtless was afraid to speak
after this plain blunt manner, as a Tyro does, for
fear of exposing the Secret too plain; but 'tis
clear, that this was the meaning, because nothing is
so hidden in Nature as the Source of this Universal
Fountain, and nothing in Nature has Power to reduce
Bodies by Symphony or consenting Harmony but what
does arise here from; for in these latter Words he
also describes the Nature and Quality of the Matter
to have two Faces; for that without there had been a
Composition, the word Harmony needed not to have
been used; so that in these Words abundance of
Matter is couched in little room, every word being a
full Sentence; how lightly or slightly soever the
Reader may pass them over: 'Tis true Starkey does
very Learnedly strike the Mark in his Exposition
upon them, yet his Comment is so wisely Regulated as
to be kept as obscure, as the Text itself; by this
means and method I find, that the Basis of this
Liquor was by this Philosopher designed to remain a
Secret to the end of time: Therefore or the Benefit
of the true desires of Art, I shall deliver the
Subject Matter of this Immortal Liquor with much
clearness and Candidness, yet hope, that my Stile
and Words will be such, as to cloath it by such a
Medium, as that it may give Light to the chosen Sons
of Wisdom, yet at the same time cast a Mist before
the Eyes of the Unworthy: This Method being
agreeable to divine Wisdom its self; for we find,
that what was a Light to the Hosts to the Children
of Israel, was a Cloud of Darkness of that of the
Egyptians.

These Things being Stated, I shall now come to the
Consideration and Illustration of that Subject
Matter they Point forth, viz, the hidden Fountain
must be of a double Nature, or what if I should say
it must be a Body of Salt appearing under two Faces,
which being united makes Symphony or Consenting
Harmony; the Reason of which is shews; for then it
is a Liquor of that excellent Purity, as to admit of
no division of Parts; therefore as I said before,
Labour couched in Helmont's Words: The Business of
Nature in affording such an universal Fountain to
the Artist, that is the Basis of the said Immortal
Liquor; and the Business of Art is to know how to
make it Corporal, and when so Corporified to contain
two Faces, which Faces Philalethes figures forth by
Urine and Blood; the first Face is a Body, yet
nevertheless may be distilled into a Spirit, nay so
Homogeneous as not to leave one Grain of fixedness
or Salt behind it, which he describes to Distil over
in Veins like Spirit of Wine, and speaking very
great of its Active Qualities in dissolving Bodies;
the Query is put, whether it is not the Alkahest,
the Answer is in the Negative, saying, it could not
Subsist without Blood, and then presently comes to
the Affirmative concerning the Subject-Matters of
the Alkahest, and says, it is Contained in Blood and
Urine; these things may grabble and amuse the
Thoughts of the Unwary concerning the Reality and
Possibility of these Assertions, yet nevertheless
they are as clear and Perspicuous to the Eye of the
Wise, as the Sun in his Luster upon the Meridien;
for the Universal Spirit being Concreted becomes a
Mist, Vapor or Chaos, or rather an Unctious and
Viscous Water, which is the true Matter of all the
Ancient Philosophers, concerning which Chaos I have
written largely in Mercury's Caduce and have there

shewn, that in its Womb is contained the first
Essence of all Forms, yet unspecificated, and so
consequently it contains these two of Urine and
Blood which indeed are the Urine and Blood of the
great World, and not of Man; but more noble; which
my Eyes have seen and my hands have handled made
Corporal; therefore I would not have you spend your
time in vain, gazing on Husks or the outside Shell
of things but Press for the Kernel, or that
excellent Sweetness which is placed in the Centre of
Beings, which can't be extracted but by profound
Meditation, and hard Labours, which must be thy
Interpreters; for 'tis not requisit, that Matters
should be discovered more plainly, it is but just
and fitting that God should be the sole Dispenser of
it, till the fulness of time, when according to the
Promise, hidden things shall be made manifest even
such as have layn hidden from the Foundation of the
World: Therefore, O Son of Art! thou must pray to
God, but use the means, and put thy Hand to the
Plow, not looking back; then these Instructions will
be as Fundamental Rules to begin thy Labours by to
obtain this Noble Secret, which is not so much a
Product of Nature but of Art: For I have in these
Sheets endeavoured to clear up the Matter, so as to
qualify thee with Theory, thence to Judge of
Sophistical Authors, and the better to enable thee
to withdraw thy Mind from their Entanglement, that
thou mightst build upon that sure Rock, which will
remain in the Storm of Tryals; this I have done in
Bowels of Love as well knowing the great Grief and
Torture of Mind undergone in my unwearied Search
after this Secret, even when the true Subject Matter
was known; which said Matter is also the Matter of
the Grand Secret of the Ancients; but diversified
into different Natures by the different Operations,

and so far distinct the one from the other, that an Artist may be Master of the one, and not of the other, and therefore he that is a Compleat Master of both, is properly stiled Adeptus Duplicatus.

To the Truth of this my Affirmation, I have not only Experience, but also the Concurring Testimony of that renowned Philosopher Ludovicus de Comite who says that the Matter of the Liquor Alkahest and Philosophers Mercury do both proceed from the same Chaos, but by different Operations are brought to different Effects; therefore before thou proceed to the Preparation of this Liquor, thou must learn to understand this general Matter, Mass or rude Chaos, which is the Source or Fountain of so many Mysteries; for this Liquor does not only proceed from it, but also 'tis the Wellspring of the Mineral Life, and while this World hath a being, will be an Inexhaustible Fountain to all those Mysteries, so hiddenly delivered by the Ancients; for once again I say, that not only the Stone, the great Elixir, but also this hidden Fire does proceed therefrom: But here you must Understand the first Chaos, before the Philosophical Mercury is produced therefrom.

Therefore a Body and not Bodies must be sought for, which being found is the Centre of the Universal Influences Concentrated and the Blood of the said Body to be one in Essence with the Body, though it appears to Sight in a twofold diversity, yet distinct in Wuality[5] or Complection, but agrees so Fundamentally, which being United by the Hand of an

---

[5] As with many oddly spelled words, this exactly matches the R.A.M.S. version, yet its meaning is unclear. -PNW

Artist will make the Symphony or Consenting Harmony before spoken of; for in this Case it may be said of it, as in another place is spoken of-the Mercury of Philosophers, that which is above is as that which is beneath, and so Vice Versa; for that the Essence and Life of the Blood can't be obtained without the Fermentative Spirit of the Earth, or Saturn's Urine; neither can this Spirit of the Earth be Homogeneous and Immortal, without it extracts the Life of the Blood: George Starkey a Disciple of Nature does in his Treatise of the said Liquor in a Parabolical way deliver himself concerning these two Faces, thus, that most Acute, Subtil and Penetrating Spirit of Mans Urine by the help of another Medium, not of divers Ferment to its self, but Centrally one with it, must be United with an Acid, not Corrosive, sed Naturae suae Gratissimum this Acid must be equally Volatile with the Salt of Urine before it can be Married or United Intimately with it; then by often Circulations it attains that height of Purityto be Entitled Ens Salium Summum alium et Felicissium, Now that which is Centrally one with this Philosophical Urine is Blood; for the Blood is the Universal Form, as the Body is the Universal Matter, but these being United by Force is called a Violent way; for 'tis a different thing to sow Gold in his own Matrix of Universal Mercury, and so ferment it and bring it into Spirit; for then it becomes unfit for the work of Multiplication, the Seminal Virtue being then totally Destroyed and Annihilated, which is the very Matter and Case of the Difference of the Mercury of Philosophers and Alkahest.

I have shewn you not only the Matter, but also the Manner and Apparition of the Matter in the Hand of

the Artist; I shall now come to shew you the Nature
and Internal Property of the same, when the
Knowledge thereof is obtained: I say 'tis wholly of
a Saline Nature, which is a middle property held up
in the Arms of Nature, and is contradistinct to
either Acid or Alkali so that neither of those, as
already shewn in the former, have any right to be
the Matter or Foundation of this Dissolvent, but
this Saline Quality is the Central one, so that
consequently this pure Spirit hath some Garment or a
Shell, by which it is Covered and in which it is
hatched and brought to Maturation; and to speak
plain, Candidly and Honestly 'tis a Combust Sulphur,
so wholy Combust, that the Spirit being drawn from
the Earth the Faeces will burn without the least
Smoak, which shews that there is neither any
Mercurial or Saline part remaining; this is a
Reduction of the Pure from the Impure, or a Clean
from an Unclean, by the Serpents devouring himself,
and then renovating into that, over which Death has
no Power: Observe, he first begins by biting his own
Tail, and so by Degrees devours himself, and last of
all his Head, which shews that the Earth or Tail is
first to be Dissolved, which then Dissolves -the
Head or Blood; and that these are the two Principles
spoken of by Philalathes is very plain; for in Pag.
25. of his Secrets Revealed, he calls it the first
Ens of Salts, saying the true Philosophers rejected
all Salts, one Salt only excepted, which is the
first Ens of all Salts, which dissolves all Metals,
and by the same Work Coagulates Mercury; but this is
not done but by a Violent way and therefore that
kind of Agent is again separated both entirely in
its weight and Vertue from the thing it is put to:

And in his Exposition upon Ripley's Epistle, speaking of the Separation of the Sulphur from the Mercury of Bodies, he says, 'tis performed by the help of a Liquor drawn from the first Ens of Salts; and Helmont and Starkey say as much calling it Ens Salium, Summum Salium Felicissimum which is the very same, that I do here assert of it: What shall I say, must I in every word Transgress the Silence of Pythagoras ; No: Be thankful for this, for it had never come to thy Hand, had I not made a Solemn Resolution in the time of my hard Labours, Sweats and Agony of Body and Spirit, that if ever the Almighty Being should bless me with the Knowledge of this Liquor, I would then deliver it so Candid, as that my Writings should be a sure Landmark to the undaunted Coaster in his Intended Voyage to the Haven of Rest; which promise I have here fully Accomplished in shewing the Universal Source or Fountain, from whence this Liquor and the great Elixer doe arise, is one, so that more needs not to be said as to this Point, therefore shall Conclude this Chapter.

CHAPTER III

THE TRUE WAY AND MANNER OF PREPARING

THE LIQUOR ALKAHEST

The Mistakes and also the true Matter being shewn, I
shall now come to shew the true manner of the
Preparation of this great Dissolvent which is very
difficult; for as Philalethes in his Exposition on
Ripleys Gates, Pag. 279. says. the Liquor Alkahest
is 100 times more difficult to Prepare than the
great Elixir; and that upon good Ground has this
worthy Author thus delivered it; for the Elixir is a
work of Nature, and the Mercurial Power doth Purge
off the Dregs Naturally, and it is called Elixir, so
long as it is Water, for as Count Trevisan says,
Azoth is drawn out of the Elixir, as Oyl out of
Water; therefore as the Elixir is Natural, the
Liquor Alkahest is Artificial, and as Ludovicus de
Comit, says, very difficult to be searched into; for
it may be variously thought of, being Artificial;
for the Subject as it tends to Generation and
Corruption in order to a more Excellent Birth, is
then vile and mutable, Proteus like puts on all
Shapes; and what we search for must be pure and
clear, and above all things Immutable; so that here
is Chymical Faith required to believe before-hand
and after sight will Astonish Reason to Contemplate
it, crying out with an Holy Admiration, O Lord! how
wonderful art thou in all thy Works.

Ready! If you would Consider the work of Creation
'twas the very same; for out of the rude Mass or

Chaos was not only Produced the most despicable
Object we behold, but also the most Glorious
Creature that ever was Created, not only Paradise,
but also the Transcendent Glorious Angels, so that
from this the difficulties do arise, how to proceed
by Art in separation of this Chaos, as also the due
way and manner of Joyning due Agents and Patients by
the separating of things adjoyned, for 'tis not
sufficient to understand the Aqua Vitae of the Wise,
but you must also know, how to preserve it from its
Compeer or Water; for being separated from the
strict Tye it had in the Elements, it would rather
pass over than come again to Coagulation: Again,
'tis easy to be destroyed, if you take it unripe, by
the violent Fire of Separation:

These Difficulties did so Amuse and Puzzle me as to
keep me back from the Possession of the said Liquor
many Years, which Knowledge I then vained abundantly
more than the Possession of great Treasures; but
however blessed be God I have great Reason to say,
that one Secret seldom discovers itself alone, for
the Knowledge of one is a large step of Entrance
into the other; for that in this Subject both of
them lye, invisibly hidden, as it were, under the
strong folds of a Mineral Ens ,which the Industrious
hand must labour to make Manifest; which being
Effected, his time will be little enough, to
Contemplate and Admire at the sight of the Operation
its wonderful Effects.

Observe, in the Manifestation 'tis reduced to the
smallest Atoms immaginable, through which it arises
to the Eminent Dignity spoken of; for as Helmont and

100

Starkey have it, Ad minimos reductus Atomos in Natura possibiles, and c. dignius de Corpus non reperiens cui Nuberet; This *Latex*, which is Vile and Contemptible, is advanced to the transcendent height of Purity and Perfection, which Words, says the Latter, are soon said, but not so soon understood, and hardest of all to be done; which is the Reason of the many difficulties mentioned in this Chapter, concerning its Preparation.

'Tis true, this Operation is in few Words taught by Paracelsus, where he says in his Treatise, De viribus Membrorum Cap. de Hepate, the Process of the Alkahest is (Ut a Coagulatione sua resolvatur, ac deinde Coaguletur in Formam transmutatum, sicut Processus Coagulandi et Resolvendi docet,) Which short Process is the greatest Light that acute Philosopher gives concerning this Mystery; no marvel then if the Doctrine of its Preparation remains so obscure in the World; for Starkey allows, that Helmont's Doctrine is equally as obscure, as that of Paracelsus, and I say that Starkey's is as obscure as either of them, and indeed that of Ludovicus de Comit. not much clearer; for that of Solution and Intervening Coagulation is the greatest Light, that any of the Philosophers have given concerning the Preparation of this Liquor; for there is this Reason of such an Obscuration, the Process by them given is General, and common and alike to most or all Chimycal Maisteries; but more especially to the two Grand Arcana's, viz the Liquor Alkahest and the Philosophers Elixir.

But the manner of Solution and Coagulation is quite different, the one is Natural, as already said, the other Artificial, and therefore very difficult, because it is not easily searched out, and that it may be variously conceived of; but that which has been hitherto known and demonstrated by all true Artists is Solution and Coagulation; we shall a little consider the difference and manner of this Solution in both these Magisteries.

The Dissolution requisit in order to obtain the Alkahest is a dissolving of the Body into a Spirit that will never Coagulate into a Body again, but the Dissolution in order obtain the Mercury of Philosophers is a Dissolution, wherein the Essence of the Body is so Congealed, as to become a Ferment to the Mercury, to Congeal upon the Body again; for as the Worthy Trevisan says concerning the Preparation of the Mercury of Philosophers, the same Matter must abide that the same Form may follow, and that nothing is to be added to it, nor taken from it, but a Superfluous Phlegm and red Earth; for when Bodies are to be Renovated it must be done by things in kind; therefore Trevisan makes it a great Error to alter Mercury from its Metallic Species; we may ground, that the Great Work is performed by a dissolution of the Body and Congelation of the Spirit, but the work of the Circulated Salt is a Solution into Secondary Principles, but not into Elements; for nothing of Profit can be expected from thence, yet this Solution ought not to be into every distinct Principle, but into one Saline Liquor, Homogeneous and Immortal. Here this Body of two Faces, or that of old Saturn's Urine and the Blood of the Great World are reduced to one, and that you may know it when so reduced, it is a Fire, yet in form of Water; 'tis an Air, yet Condensed; 'tis no Corrosive, yet the most sharp and perpetual Corrosive; 'tis not Medicinal yet the Crown of all true Medicine, being a Cleanser and Purifier in Nature, a Destroyer and Conqueror of Bodies; 'tis called the Fire of Hell, because the Spirit that comes from the Centre is United to the Blood without the Intermediation of the Heavenly Rays of Mercury , yet Acts with that Purity, as that it finds no Body more noble than its self to joyn withal, therefore is not Commiscible with any Ferment, and so not capable of Transmutation.

These Things being considered I shall now come to give you a short Scheme of what Helmont says concerning this Liquor, the first is what the Artist desires, and is Comprehended in these Words, Art is Solicitous in finding out a Body, which may play with us in such a Symphony or consenting Harmony by reason of its exquisite Purity, that no Corruptive Principle can find in it any Heterogeneities, by which to work in it a Dissipation of Parts: This is the Sum of what the Artist would attain, and is the Chief of all which can be by Art sought for. The Second is, what Art by Industry doth find, Comprehended in these Words, Religion then stood amazed, the Latex being found, which being reduced to the smallest Atoms possible in Nature despises the Wedlock of every Ferment, in vain therefore is its Transmutation sought for, not finding a Body more worthy than itself whereunto it may be joyned. The Third discovers the Anomaly of this Production, in these Words; But the Art or Labour of Philosophy hath brought forth an Anomalous Product in Nature, which took its being without Mixture of any Ferment, divers or Heterogeneous to its self and the fourth contains a short Adumbration of the Process, the Serpent bit himself, revived from the Poyson into a pure Essence, over which Death hath no Power. All which to a Son of Wisdom I have with much Candidness already explained; but for the benefit of Tyro's I shall now come to give a farther illustration by way of Mechanic Demonstration.

If Art will from the Latex obtain a Body, it must be begun with Coagulation, and with such Magnetical Earth, as Attracts the Celestial Raies and Universal Spirit, and Concentrates them in the Centre, for that in the Centre the pure Parts of the Starry Fire is digested, and in the Centre all the Influences meet, and from the Centre does the living Ferch proceed; for 'tis the Central Archaeus that Sublimes the Mineral Vapour or those pure Spirits that are of a dissolving Nature; this is the Body which Art is Industrious about and desirous to know, even the Body of the Universal Salt and Sulphur of the great World; for in the Mechanical Demonstration from thence proceeds such a Spirit which in all Nature has not its Compeer; but before 'tis brought to that Harmony, as to admit of no Dissipation of parts, it must have time to maturate, and to form to its self some pure Garments or a Compleat Coagulation, which is done by Successive Animations, so as to bring the Spirit of the Earth to permanency before it is taken out of its Nest, and that is done by successive Retrogradations, or repeated Coagulations to bring this Transverse Work of the Earth the nearer to its purity, for then only it is, that it admits no dissipation of Parts.

The Body being found and thus purified, and its Spirit produced from the small Invisible Putrified Atoms of the same doth cause a Religious Astonishment; that from so dry a Body as the Earth should proceed the Central Latex or the most hidden Rivulet of the great Ocean its self, nothing in the World being so pure it despises to Contract Wedlock with every specificated Form whatsoever, and so its Transmutation is impossible, and indeed the oftner this Universal Spirit passes through the Entrals of the Elements, the greater is its Purity and the fitter for Action; for on the other hand Matter could not Subsist, but by the continual additional Rays of its Universality; therefore, Reader, Retain thy Amusement for a time, and thou shalt hear the Philosophical Trumpet calling thee to behold a wonderful rarity, even the Indian Brachman's Famous Water Works, Contradistinct to all others, as it is a well compacted Body of Fire burning in Water, and in full Lustre and not extinguished; for in the Decoction the Blood and Urine is Centrally one, but in the Coagulation they appear under two Faces; Philalethes says as much, speaking of the next Matter of the Alkahest, says 'tis a Salt and the Fire surrounds the Salt, and the Water swallows up the Fire, and yet overcomes it not, and so is made the Philosophers Fire, of which they speak, the Vulgar burn with Fire, and we with Water: it being so transcendently pure then scorns to be joyned with any Compeer, so admits of no Wedlock.

Herein Consists the Excellency of this Liquor that Art and Labour conspire together to produce this Ens or being without any mixture of any ferment Diverse or Heterogeneous to its self; for indeed its Composition is wholly from Universal Principles, or Virgin Elements, and Invisible ones too, for the Earth and the Water of this Fountain is as Invisible to the Eye of the Vulgar as the Fire and Air, but being made Manifest to the Artist do Conspire together to bring forth one Anomalous Birth; therefore be assured, from these Words this Great Philosopher would point forth, that this Liquor is not Compounded of two things of different Natures, for then it would be Subject to Death, but of one thing alone even the most Universal Salt of Nature, which is divided into two, and returns to one again, so consequently is one in Nature and Essence.

But now the Adumbration of the Process doth also shew the Universality and Uncompoundedness of this Ens, it is represented by a Serpent biting himself and reviving from the Poyson into a pure Essence, over which Death has no Power; observe the Ancient Egyptians always by a Serpent understood this Universal Spirit; because of the Subtilty of its Parts, and that Creature of all others is the most Subtil, and therefore well may it figure forth by the holding its Tail in its Mouth Natures Circulation from one Universal Element to another, till all the Elementary Qualities are unbanded, and the pure Essence of all extracted, which moves upon an Immortal Hinge and therefore will admit of no Dissipation of Parts or Evaporation of Moisture; for as it is Homogene, 'tis of an equal Volatility, and being the Celestial Fire made Terrestial by its own Power can't be destroyed by any specificated Body whatever, therefore well might Helmont say, as there is but one Fire in the World, so there is but one

Liquor, none other partaking in Quality with it, as the Adepti do know and can testifie: Which Liquor is that which will, Sampson like, make sport for the Artist, and do more than ever the Lords of the Philistines could have expected from him, plucking down and destroying as he did not Houses, but the most Compact and Solid Bodies of Minerals, and like a Valiant Conqueror maintains his Ground against all opposing Enemies: but few are the Kings, Lords or Nobles, that have been so happy to see the Battle fought by this Anomalous Combatant, not only for the Reasons afore rendered, but also for the tediousness of the Preparation, which we shall now come to speak of and so Conclude this Chapter.

As to the time of preparing this Liquor 'tis long and tedious, which Helmont also Complains of, but Starkey explains this to be more upon the Account of the Stink in the first Preparation, than of the tediousness or length of time; for he Limits it to a few certain Days, which I know is impossible to be Effected, unless he begins to Calculate from the time that the Body is ripe and fit to be broke by Violence, and to be distilled over into a Spirit, then to Unite the Blood or other Face may be accomplished in his time: But for my part I do comprehend from Helmont what Experience shews, viz that the time is long and tedious, and attended with many difficulties, which Helmont also knew, or else he would not a Complained so much for the loss of his Bottle; Crying out, O that I had removed my Receiver; from whence we may readily Conceive, that he suffered Loss: Ludovicus de Comitibus puts the Question beyond doubt concerning the tediousness of the time where speaks of the Reduction and

Retrogradation saying, it can't be performed by common Labour; but requires both Art and Time, which, he says, is long and therefore whoever thinks to obtain it by Violence and in a short time shall find himself much deceived, for he can never bring it to any final Complement, and consequently will never be able to know what Vertue 'tis Empowered withal, even that Fiery and Vital one; for that it is destroyed by force by his Compeer, which Helmont and Starkey do allow, but tell not what that Compeer is; but that excellent Son of Art Ludovicus aforesaid, says 'tis Water, therefore I'll attribute to him the Praise, his Writings giving me the first Light of discovery, what this Compeer was.

O Reader, thou must of necessity allow that it is a time of Tediousness as well as difficulty to Concentrate the Benignant Spirit of the World, to make such a strong Sulphureous and Saline Liquor of it, as will dissolve the hardest Metals, even those, that oppose Common Fire, are by this Liquor radically opened, it being as we may call it, the very Essence of the Elements Heavenly and Earthly; and what Typifies the Fire of the last Judgment, which is permanent over the Elements in a Quintessential Nature, so that the Degrees of its purity are not to be wondered at; the Artist hath great Cause to Bless and Praise the Lord his God, who hath made him an Instrument to produce a clean thing out of an unclean, which that Good Man Job so much Questioned saying, who can bring a Clean thing out of an Unclean, surely none but God alone; so that we see the best of Men speak but according to that knowledge, which the Almighty thinks fit to reveal unto them; for he disposes of Knowledge as of

Rivers, communicating it for the use of all:
Therefore Paul's Advice was Sound and Candid, Judge
not ignorantly of things thou understandest not: For
indeed if we look upon the thing aright 'tis
properly the Work of God; for as Trevisan says
concerning the Exaltation of the Work in the great
Elixir, 'tis done Christi Gratia, Importing that Man
can't alter the ordained Course of Nature, but as an
Instrument in the Hand of God stands still to see a
mighty deliverance; but we are speaking of that
where Art must lend her help, because Nature is
altered from her usual Course, and a Clean thing is
also required; but this is a Talent not Committed to
everyone's Trust.

Now this clean thing can never be Produced but by a
Radical Union of the foresaid Principles, not only
by a bare Association or Apposition of Parts, so
that the same may be said of this, as is of that
Union of Sulphur and Mercury in the Great Work, viz,
they can never more be separated, neither in Love
nor Woe, this Radical Union is as Principally
required in this Liquor, as it is in Azoth, which is
a Volatile Tender Spirit for Whitening Eaton: Now
this being separated from many Heterogenities hath
no Eminent smell, but is a Ponderous, subtil Liquor,
which will not stifl over but in a considerable
Degree of Heat in Sand, viz, the third Degree, and
admits of its Flegm to be distilled off first, as
other Ponderous Spirits do: Philalethes speaking of
the Substance and Preparation of this Liquor
delivers himself thus; 'Tis a Noble Circulated Salt
prepared with wonderful Art, till it answers the
desires of an Ingenious Artist; yet 'tis not any
Corporeal Salt made liquid by a bare Solution, but

is a Saline Spirit, which Heat cannot Coagulate by evaporation of the Moisture, but is of a Spiritual Uniform Substance, Volatile; which in a gentle Heat will Distil over, leaving nothing behind; that is to be understood in a Requisit Heat of Sand; so is there an Exaltation made far above what Nature was ever able to perform.

Thus I have delivered the difficulties and also touched at the Fire and given you the right way of its Preparation from Point to Point, I have declared the Truth without Defect or Ambiguity of Words; and have as formerly mentioned, shewes you that no strange Ferments are used, the Principles being Centrally contained in the Original Chaos, which being separated and brought again to an Indisoluble Union, is, the Serpents devouring his own Tail and so renovating into that, upon which Death can have no Power: but this cannot be performed, but by the help of Fires of divers sorts, Convenient Vessels, fit Furnaces and Glasses, and a considerable time to boot, without all which it will be impossible for the Artist to obtain his desired end, being as I have delivered, much easier to know the Matter, than to find out the true manner of its Preparation, which is chiefly and principally to be sought for at the Hand of the Almighty; these are Secrets which belong to the Divine Treasury, and therefore the Aid or free leave of the Triune Power must be implored, to open the Door of Entrance, which may otherwise forever remain shut to thee.

Yet I have not been wanting in these Sheets to Communicate my Experience, and Candidly to shew the

Preparation of this Immortal Liquor, so far as was lawful for me without exposing it to the Hands of debauched Persons and Impostors, and he that can't gather it from what is here delivered will scarcely obtain it from the Voluminous Circumlocutions of other Writers; who have so Intermixt the Preparation and entangled it with the Philosophical Mercury, that the Artist stands in need of Ariadne's Clew to lead him out of that Labyrinth; in which Maze that the Ingenious may be no longer bewildered, I shall in the following Chapter distinctly and clearly discover the difference between the Liquor Alkahest and Mercury of the Philosophers.

CHAPTER IV

THE DIFFERENCE BETWEEN THIS LIQUOR AND THE

MERCURY OF THE PHILOSOPHERS

The Labour of the Candid and honest hearted is to
untie those difficult Knots, which the envious have
always been endeavouring to tye, and to bring the
Industrious out of that Labyrinth, where they have
been entangled and bewildered, so as to lose the
Right Path, and for this end I am willing to lend my
Hand to conduct the searcher through this Wood where
many an honest hearted and laborious Man I am well
satisfied have lost their Way, as not being able to
distinguish the different Path of the Liquor
Alkahest and Mercury of Philosophers, designing this
Chapter as a Plain and knowing Pilot in this Case.

I hope that nothing but Ignorance itself will
question the Verity of what I have here delivered
concerning the Foundation of the Alkahest and
Mercury of Philosopher to be one, seeing I have on
my side not only Experience, but also the Testimony
of worthy Sons of Art, that they do both proceed
from the first Chaos, before Art hath undertaken to
work upon it: But here the difference comes, one is
prepared in a way agreeable to Nature, the other
Artificial; and consequently really divested from
the Generative Power, being drawn beyond the
Predestination of its Natural Seed; the exact
Example of which may be seen in a Grain of Wheat,
when 'tis sown in its proper Matrix, in order to
Multiplication by Generation; or when it is

Artificially Prepared and Fermented, and so drawn into Spirit, in which Work the Seminal and Generative Virtue is wholly destroyed: For here there is made an Artificial Solution of the Seed not into Elements but Secondary Principles; and by this violent way of Dissolution 'tis divested of its Metallic Seed, and Consequently made unfit for the Act of Generation, as was shewed just now in the Example of the Grain of Wheat: so by consequence must bring a considerable difference at their Respective Ends.

Yet nevertheless both these as they arise from one Universal Fountain there may be some likeness in them and for this Reason the Description does in some Sense resemble both the one and the other, that few have been able to distinguish the true difference, and the more by the shifting Speeches of Writers, who confound the one with the other, that so the Artist may be easily entangled, because they have not so much as differenced them in Name, Nature or Operation; for Van Helmont says, that the Liquor Alkahest dissolves every Visible and Tangible Matter into the first Ens, preserving its Power, which Words preserving its Power is also attributed to the. Mercury of Philosophers; other Philosophers say, 'tis a fiery Water, and Lightsom, and Turba Philos. and Senior say, our Water is a Fire, and stronger than any Fire for it reduces the Body of Gold into a meer Spirit, which the Common Fire could never be able to do; the like also says Artephius; the very same thing is by others attributed to the Alkahest: Helmont says that as there is but one Fire in the World, so there is but one Liquor in the World, no other partaking in Quality with it; and

113

Geber says, the most high hath given us the Knowledge of this Water, which lights the Candle gives Light to Houses and yields abundance of Riches: It would be too tedious to enumerate the Parallels of this kind concerning the Alkahest and Mercury of Philosopher: so that 'tis very difficult for the unskillful and unwary to distinguish their true difference, which is mostly to be Comprehended from those Words, where 'tis said, the one is a work of Nature, the other of Art; so are they different in appearance, for as a late Author says, that I may prevent a Common Error, viz, the confounding our Natural Dissolvent with our Circulated Salt or Alkahest some Ignorant Boasters who neither know the one, nor the other, having taught that they are both the same, I shall so far shew the Difference, that no Tyro but may effectually distinguish them in his Theory. Know therefore, and Note well this short distinction, there is no Affinity between them either in Matter or Operation: They differ in Matter, as much as one Species doth from another, the one being Metalline the other Saline: They differ in their Operations, as much as Love and Wrath; the one in Love Preserving, the other in Wrath Destroying, Life and Motion.

This Author by his good leave speaks right in the Operation, but strains the String too far concerning the Matter, as too many Reformers do, and so cause Errors on the other Hand, equal to those they would Reform, thereby making many to grope for the Door of Entrance, or middle way, which leads directly to the Path of Truth: For be sure as they proceed from one Matter, both Universal, there is something of Assimilation in them; for as much as they are both performed by way of Solution and Coagulation, both tedious and difficult in searching out; and the Subject Matter so far exalted from its former State, as that it becomes a Work of Wonder; and for certain they must have something of likeness, or else those Artists were very Ignorant, who gave them one Denomination, calling them by the like Name, as Fiery Water and Watery Fire, Immortal and Homogeneous Essences, Alkahest, which is all Ghost or Spirit, the first Ens of Salts, and have attributed Supernatural Vertue to both; and from these and such like Universal Terms and Names, 'tis very easy for the Searcher to be deceived.

Wherefore I shall now come to give you a clear and general Account, wherein they agree, and wherein they disagree, and then shew you the Reason, why they are thus described, then give you the true and proper signification of the word Alkahest, and why Helmont gave the Liquor this Name; for I have taken some pains after the Inquiry thereof, so that I am able to render a Satisfactory Reason:

First of all I shall Instance in some particulars, wherein they agree: First, the Mercury of the Philosophers and this Circulated Salt agrees, in that they are both Universal, one for the Graduation and Exaltation of Metals, the other for Dissolving all Bodies: Secondly, They agree in this, that as one preserved the Seed in order to Multiplication by Generation, so the other preserves the Crasia and Medicinal Vertue of Species in order to healing; for in the Dissolution it admits of Nothing to fly away in Fume: Thirdly, They agree in this, the one is the

Emblem of Man's Regeneration and eternal Salvation, the other of Man's Dissolution and Destruction; for in the Preparation they are both to be seen: Fourthly, They both agree in the Penetration of Bodies; the one enters to the very Central Life of them in order to the Multiplication, the other pierces to their very Centre in order to their Separation and Division; for it separates between their Central Mercury and Sulphur: Fifthly, They agree in the Matter and Manner of Preparation, as to the Matter they both Proceed from the first Ens of Salts, and as to the manner, 'tis by Solution and Intervening Coagulation, till brought to an exalted Perfection: Sixthly and Lastly, They agree in that they are both made from the Universal Chaos, as also in the manner of their Composition; for the Mercury of the Philosophers is Compounded of Sulphur and Mercury; but the Liquor Alkahest of Salt and Fire and Blood; and both brought to such an Indissoluble Bond of Love and Unity, as never to be separated either in Love or Woe; both Homogeneous and Immortal, and both Universal Dissolvents: Having shewn wherein they agree, I shall now come to speak of that wherein they disagree.

First, They differ in this, whereas the Work of the Philosophers Mercury is purely Natural, so the Process of the Immortal Dissolvent is meerly Artificial: For as in this Work the Sulphur or Gold is exalted to the highest Pitch and Degree of Perfection so in the preparation of the great Hilech of Paracelsus, it is reduced from a Natural to a Contranatural State. Secondly, They disagree in this, where the Mercury of the Philosophers is an Homogeneous Metallic Ens Co-essential in all its parts, true Mercury, of a middle Substance clear like pure Silver, being bright Celestial and Shining, and not so Essential to anything as Gold, it being its Universal Mother does radically congeal upon it; therefore as Trevisan says, no Menstruum is profitable in the Philosophic Work, but that which dissolves the Body in a Generative way, and then recongeals upon the Body dissolved, so the Philosophers Solution of the Body is a Congelation of the Spirit; and upon this Account they have rejected all those Solutions, as Sophistical, where the Dissolvent and Dissolved remain not Permanent together: Whereas the Alkahest or Sal Circulatum is a Saline Liquor, and therefore by Paracelsus sometimes called the Liquor of Salts and doth Dissolve Bodies, but remains not with them, being as easy separable from them, as the Spirit of Sulphuris from Oyl. Thirdly, There is a Disagreement between the Mercury of the Philosophers and Liquor Alkahest in the manner of their Operation and Action on Bodies, for the Mercury dissolves Gold and all Precious Stones and Pearls by way of Generation, and Exaltation, for the Life and Vertue is Multiplied, and they may be reduced to their first Form in greater Vertue and Beauty, and of more Value to the Metallurgist and Jeweler, but the Liquor Alkahest dissolves not only Gold, but also all the other Metals, by way of Destruction, so that the Generative Virtue is defaced and wholly obliterated, and in this Reduction into their first Matter it gives a certain Testimony of their Diversity, as Metals into Sulphur and Mercury, Stones into a Saline Liquor, and Pearles into a Milky Juice. Fourthly, They disagree in this, the Mercury of Philosophers at the end of its Preparation will become fixt and Permanent abiding all the fiery

Tryals, in form of a Calx, yet as fusible as Wax Penetrating. Mercury, and other Volatile Bodies before their Flight and fixes them; whereas the Liquor Alkahest at its respective end of Preparation is a Ponderous Saline Liquor in form of Water, which will moysten the Hand and everything else, and as it is wholy Saline and Volatile 'twill not endure the Fire, but will remain in its form Distilling over in a Saline Liquor being altogether Incapable of Coagulation, and by that means dissolves all fixt Bodies whatsoever, not into Elements but into more simple Parts. Fifthly, Their Difference consists in this, whereas the Mercury of Philosophers is made by a remiss Fire of Generation, even the Aerial Life and Lunar Fire being the Medium in perfecting it by gentle Decoction from Point to Point, which Regimen of the Fire has been carefully bid by all Artists, in that 'tis called the Vessel of Nature, or Mercurial Vessel, Pondus Naturae; Whereas the Liquor Alkahest is made by the most violent Fire of separation, for the Spirit is by Violence not only Distilled from the Earth in Fiery form; but that is United to the Blood, which produces that Hellish Fire that brings all Imperfect Metals to a greater Imperfectness, though notwithstanding it makes them the more Powerful and Efficacious for the expelling and rooting out of Diseases and Infirmities, for being brought to their first Ens they dissolve and circulate with our Juyces, as being then thin and Spirituous, and so perform that in the curing Diseases, which in their hard and gross Natures could never be expected from them. Sixthly and Lastly, The Philosophers Mercury and Liquor Alkahest differ in this, the one may be brought to an Universal Medicine, the other has no Medicinal Vertues in it: For as Philalethes says, this Mercury thus renovate or new born, may by the Philosopher be diversely handled; for he may take it from the Fire, and Circulate and Cohobate this Mercury by a Peculiar Operation, which is partly Mechanical, till he have a most admirable, pure, subtil Spirit, in which he may dissolve Pearls and all Gems, and Multiply them or his Red Stone, before it be united with a Metal in Projection, for the making of Aurum Potabile: and in this Mercury, thus Circulated, is doubtless the Mystery of the never fading Light, which I have actually seen, but yet not Practically made. In a word, everyone who hath this exuberate Mercury bath indeed at Command the subject of Wonders, which he may employ himself in many ways, both admirably and pleasantly. And certainly, he that hath this, needs no Information from another; himself now standing in the Center, he may easily view the Circumference, and then Operation will be, next the Spirit of God the best Guide: So that the Mercury of Philosopher, being brought to fixity, may be made an Universal Medicine, for the Curing all Diseases, and Renovating and Restoring to Youthful strength and Vigour; whereas the Liquor Alkahest, be it never so highly multiplied or Exalted, cannot properly be said to be a Medicine, but a Menstruum, which is a Proper help or Medium to prepare Medicines by, and in itself still remains unchangeable, being as Starkey says, endued with these Qualities; Viz 'Tis a Ponderous Liquor, being indeed all Salt, without any Watery Phlegm; it is all Volatile being wholly a Spirit, without any Corporeity left in it, of no eminent Odour, for all things which send out an odour considerable, are for the most part of a very Volatile Nature, or consist of many Heterogeneities. It is not therefore Volatile after the manner of Spirit of Wine, Urine, or the like, which fly with the smallest degree of heat, but (like unto a ponderous Spirit, which yeilds its Phlegm in the first place) this when it bath dissolved any Vegetable Concrete, and made it Volatile, will suffer the same by a gentle heat of Balneum Maria, to be all separated from its self, and c.

From what has been here said, concerning the Agreement and Disagreement of these two, I hope, the diligent Inquirer after Art will receive good satisfaction, and for the future be freed from those doubts and Errors, which might before occasion much Trouble and Perplexity of mind: This was the end I proposed to myself, throughout the whole of my discourse; this therefore may suffice as to this Point; I shall now proceed to speak of the Proper Names of this Dissolvent.

But by the way (Reader) observe, that the Invention of this Liquor, in these Parts of the World is owing to Paracelsus; thus Philalethes, and also adds, that among the Moors and Arabians, it hath been, and is at this day, commonly known to the Acuter sort of Chymists, then consequently we must depend, that Paracelsus did give the most significant and Proper Names to it, and 'tis plain from Helmonts own Writings, that he diligently Studied and Traced his Works, and at length through Labour came to understand them, and amongst other things, obtained the Knowledge of such a dissolving Menstruum, as Paracelsus often writes of; and seeing this Liquor to contain an homogeneous Nature, spiritually acting, and after almost innumerable Actions still remaining the same, (Spirits being immortal) and this Liquor proving so, be therefore not improperly called it, Alkahest; although, as I shall shew by and by, this Name doth more properly belong t0 the Mercury of Philosopher, and that, this was the design of Paracelsus in it; however by the way, I shall examine the derivation or Root, of this Word, which is from the Belgic or rather High Dutch. Language; in Holland or Flanders, where Van Helmont lived, (Geest) is as much as to say in English (Spirit) and in the German Tongue, 'tis much higher and Guttural, being expressed (Alchahest) which signifies (All Spirits or all Spiritual:) which Word (Alcahest) Paracelsus makes mention of in the Tenth Book of his Archidoxes. Chap. 6th. where treating of the Virtue of the Members, says that the Liquor Alchahest, has a great power of Conserving and Comforting the Liver, and consequently of Preserving it from the Dropsie, and all such as arise from the defect of the Liver, and if the Liver is dissolved or broken, it stands in the Place of a new Liver: The Process thereof is this, it must be resolved from its Coagulated state, and Coagulated again into a Transmuted Form, as the Process, of Coagulation and Dissolution Teaches. This Passage is the only Place, wherein Paracelsus has made use of this Name, it being not to be found elsewhere in all his Writings, so that 'tis plain to us that Belmont has borrowed this Name from him, therefore we must according to Reason and Experience consider, whether Paracelsus meant this Liquor or not; because the Process set down, viz, Solution and Coagulation is alike and Common, (as hath been already touched at,) not only to the Preparation of both these but likewise to most Chymical Magisteries.

Now the Liver is the Fountain of the Blood, and is the seat of Life next the heart, the Blood being there Prepared for a further Elaboration and Purification, in order to give the Body, its Nourishment for the Production of Seed, and Consequently for the maintaining of Life, and c. And 'tis plain by experience; that this Liquor will by greater length of time, dissolve all mixt Beings by its Active, Thin, Spirituous Penetrative, Dissolving and Homogeneous Nature, in a Natural degree of Heat equal to that of the Liver, and separate them into their distinct Substances, suffering not anything to fly away in Air or Fume; so that to me here arises the difficulty to think, how this should work that different Effect, of healing and restoring the Liver, and not rather dissolve it, as it does other mixt beings: The doubt is beyond my reason at present to give an Answer to, and I suppose will so remain forever, for I do not so much as once intend to an Experiment, to try its Virtues in this Case; having, besides what is already offered, two Substantial Reasons against it; the First is, that this Liquor being difficult to be prepared, would be too costly to be administered by way of Medicine; for a Reasonable Practice would soon diminish a considerable Quantity, so that this great Treasure would in little time, be exhausted and come to nothing, if given by way of Medicine, whereas 'tis perpetual by way of Menstruum: The Second is, that the Philosophers give no directions for the Exhibition of the White Stone inwardly, but in Epilepsies and Palsies , and other Diseases of the Brain, which is under the dominion of the Moon, much less its White Oyl but for Externals, as Leprosies, Scabs, Virulent Ulcers, Fistulas, Cancers, Noli me Tangere, etc., and the like; how they should then dare to exhibit a Spirit so Active and Fiery, yet much more Crude than these, I know not; neither indeed can I be made to believe, that ever Helmont or any other of the Adepts , did ever once so much as make use of it by way of Medicine, and Consequently could not be this Liquor , which Paracelsus meant, where he speaks of the Cure of the Liver, but rather of the Grand Elixir.

But 'tis abundantly more probable, that they served themselves therewith in the Preparations, of Drugs and all kind of species, in order to bring them to Magisteries, Arcana's, Essences, and Quintessences, which have a superlative Virtue, especially from the Metalline and Mineral Kingdom; because what is resolved by it retains their healing Faculty; so from these Considerations I can't Conceive that Paracelsus, where he speaks of the Restoring of the Liver, that he meant the Circulatum Minus or this Liquor, so that 'tis altogether indemonstrable, that this single Dissolving Menstruum should be a safe and good Medicine, and Consequently should Cure the Dropsie as is easy to be gathered from the foregoing Words of Paracelsus, that his Alkahest really was Medicinal; for he expressly says, if the Liver were broken or destroyed, it would be in place of a new Liver; now from the foregoing Considerations, this Liquor can't be said to be a safe and good Medicine, 'tis therefore abundantly more probable that Paracelsus by the Word Alchahest meant the great Elixer, that being all Spirits, a Quintessence, divested of all the Elements, and consequently of all Earthly and Corporal Qualities; for if the Grand Elixer were not Spiritual 'twould never Transmute; for by this Spiritual Act, it works three Effects, first Penetration and Dilatation; secondly by Fermentation and Contraction; thirdly by the Acts of the two former, the combustible Sulphurs are separated, the pure ones manifested with additional Tincture and Permanency so Helmont finding his dissolving Menstruum Spiritual might easily mistake the Words of Paracelsus, and call it Alchahest, and indeed the Name is no ways Improper, although not used for this Liquor by Paracelsus , unless Paracelsus was guilty of speaking one thing and meaning another, as Helmont himself sometimes is; as I can prove from these following Words. The Liquor Alchahest (says he) Reduces all sensible and Tangible Bodies into their first matter, Preserving the Power of their Seed; which as you have all along heard it doth not; but their Medicinal Vertue; the Property of Preserving the seed belongs to Azoth or Philosophical Mercury; So that if he were not guilty here, he was for certain beside the matter; but I am apt to believe he was, seeing he has not in all his Writings given account of the Medicinal Vertue of his Alcahest, as Paracelsus does of his.

From hence it may be clearly Conceived, what I have inserted in my former Doctrine, that the Philosophers were many of them guilty of interweaving these secrets together, and calling them, by one Name, for 'tis clear, that Helmont called this dissolving Menstruum the Liquor Alchahest, yet says it preserves the seminal Vertue, when as Paracelsus by this Passage meant the Grand Elixier; and the more evident in this, in that he has given other Names to this dissolving Menstruum, and those mostly used by him are the great Hilech and Sal Circulatum; for these are generally to be traced through his Writings, and 'tis easy to be discerned, that he puts a great distinction between this dissolving Liquor and the Mercury of Philosophers; for the Liquor, he calls Circulatum Minus and Mercury of Philosophers Circulatum Majus as is plainly to be Proved from the Process given, where be saith, thou must extract the first Ens of Mercury by Spirit of Wine, and it will come over in a Liquid substance which (says he) is called by the Philosophers a most sharp Metalline Ace turn, and by us in our Archidoxes Circulatum Majus. Archidox Lib 10. Chapter the 4th.

This Distinction and Process cannot be rightly understood by any, but an Adeptus Duplicatus for to the obtaining of this Spirit of Wine the Work is one, and is Performed by the concurring help of an Assistant; otherwise 'twill be impossible to be obtained but being gotten, the difference Consists in the Forcible way of dissolving the Body and the Natural by the Spirit of Wine, to extract the first Ens of Mercury,, in which the Blood is united and Cleansed, and so brought to the gentle or Benignent Fire of Nature, which is one with Central Salt Nitre and also the Magical Sol; for it unites to the Center with a wonderful Fermentative Power: Now this Spirit of Mercury, or Mercurial Fire and Oyl, is by Artephius not improperly called the Vinegar of Mountains, and by Paracelsus the most sharp Metalline Acetum; for it performs that which common Fire could never do, vizt. dissolves the Body in Preservation of the Form, and brings it to a Spirit, to be exalted aloft in the Air, where Celestial Purity, and the strengthening Multiplicative Vertue is; that Spirit, will again return to and unite with

the Body, which Circulation is continued till the
Universal Mercury, has extracted the universal
Sulphur, and then is it truly and properly called
the Circulatum Majus, or Alchahest as thou pleaseth,
the Name being proper to the Elixer itself, as may
be plainly discerned from the foregoing Passage of
the said Author, where he tells you, that when it
has overcome its Like, it becomes a Medicine for the
Liver, excelling all other Medicines; and towards
the end Adds, Verily should the Liver itself be
broken or dissolved yet this stands in the place of
the whole Liver, no otherwise than if it had never
been broken or dissolved as aforehinted: So that the
Medicine (from the Authors own Words,) by which the
Liver is cured is no other than Mercury Prepared,
Sublimed, Vivified into a new Life; and having
passed the gates of Death comes to be united into a
twofold Life, Terrestrial and Celestial, and so
becomes that Medicinal Tincture, which is a true
Emblem of Man's Spiritual Restoration, and is in a
far higher degree of Perfection than this Circulated
Salt can be conceived to be; seeing it may be so
highly exalted, as to be brought to an Elixer of
Spirits, which in a Minute penetrates the Center of
Bodies, being a Perfect Concatenation of an
Incombustible fire and light, which will admit of an
endless Multiplication, being each time advanced in
Vertue, Power and Spirituality; so that it becomes a
Medicine not only for Man, but also for Metals,
making them both Perfect and Permanent, the which
this wrathful Liquor cannot perform.

For this great Magistery hath in it the Exalted
Vertue and Universality of Light, a Quintessence or
Medicine of the highest Purity in the three Kingdoms

of Nature, Animal, Vegetable or Mineral; therefore may be properly said to be a Medicine for the Liver, this being a Member or Part; which so much Concerns the Life of man: Now this will manifest itself here a Medicine above all Medicines in order to restore firm Sanity: And that it may be yet more plainly conceived, that Paracelsus spoke concerning the Philosophers Tincture., I will quote the Words of the famous Arabian Prince Geber in his 4th Book Chap ten 1. There is a Medicine (says he) of a twofold Nature of the third order, yet but one in Essence and manner of working (afterwards cunningly adding) there is notwithstanding an Addition of a Citrine Coloured Sulphur, which is perfected by a most clean substance of fixed Sulphur: Behold how its like is overcome after the first Preparation This plainly shews that the like, which is to be overcome, is the very same, that Paracelsus spoke of, as I have experimental Reason to believe: This from the Testimony of Donnaeus and Ludovicus de Comitus, is also confirmed to be that of the great Elixer.

Concerning its Spirituality, I shall add a Passage or two more for the greater Confirmation of what is here said: Basil Valentine, (in his last Will and Testament and Allegorical Expressions (Page. 347.) between the Holy Trinity and the Philosophers Stone,) Compares his Mercury to God the Father, as being a Spiritual Body; and the Philosophers Sulphur or Gold to God the Son, who is God and Man, which Sulphur must dye and rise again for its Brothers and Sisters sake, being then a glorious Body, redeeming and Tinging them to Eternal Life: and when these two come together saith he, they are called Mercury Duplicate; from whence proceeds our third Substance,

which is our Glorified and fixed Sol, the
Philosophers Stone, or Spiritual Essence of the
Philosophers, called the Triune Stone, proceeding
from Two, Water and Spirit, Animal and Vegetable in
the Mercury and the Mineral living Sulphur of Sol ,
which are Three, Two, and yet but One. Now observe,
this Authors Mercury Duplicate is the same (Like,)
with that of Geber and Paracelsus, which the Mercury
will overcome, and then it becomes the Medicine or
Alchahest spoken of.

Thus having given you some Account and Reason, of
the Names imposed by Authors on this Liquor, I shall
now come to give you some Reason for our Additional
Name, viz. Trifertes Sagani, which is as Proper a
Name, for this Liquor, as any given by the Adepts,
it being Spirits born in and Predominant over the
fire, nay it inhabits the fire, even that fire that
hath Power to dissolve the four Elements and Reduce
them to its own Nature of Universality. Now this
Liquor being thus Prepared is a Compleat Key to the
Medicinal Art, and doth open the Treasury of
Medicines in the three Kingdoms of Nature, in a way
succedeanous to nothing but the great Elixer.

But seeing the use of this Liquor is manifold and
various and will require a whole Chapter, I shall
omit speaking of it here and refer you to the next;
where its vertues are fully shewn; and come a little
to consider the Exercise of a laborious Searcher,
which he meets withal in his search: The first
Exercise is to come to the knowledge of a true
Subject Matter which is very difficult: for the
Philosophers Words concerning It are so obscure and

hidden, and the Matter Involved in such Tropes and Metaphors, that it requires a more than ordinary help to come to a right understanding to distinguish rightly and truly and genuinely what the Matter is, which beyond all Controle is candidly done in these Sheets.

The second Difficulty that the Labourer meets withal, is to distinguish between true Books and those which are false and Sophistically Written, which indeed is a Labyrinth, equally as difficult as the former concerning the Matter; For a false Author is like a wrong Guide upon a Journey, for if in the beginning of the same, He goes but a few steps in a wrong Path, and then follows the same, may in the Conclusion be led clear contrary to his designed end, and indeed it is of greater Consequence in search, because there is few or none to be met with in all our Course, that can direct to the right way: Now there is in Scripture a Curse pronounced on all those, who put the Blind out of his way, which Curse will take hold on all those Sophistical Writers, In that there is no Blindness greater than the Spiritual Blindness, whether in things Natural or Divine, and therefore 'tis a very great difficulty to distinguish them asunder, which being done the false are to be shunned as much as the Devil himself, who is the Author of all Imposture.

A Third Difficulty is after you distinguish Authors, to come to some knowledge and Understanding of the true, concerning the Scope and Intention of their Writing, both as to Theory and Practice; which indeed is a Difficulty surmounting the former,

former for these Reasons: The first is their Circumlocutions and large Descriptions of things when as indeed it may be comprehended in little Compass; the second is their Multiplicity of Repetitions of one and the same thing, only with some variation of Words, only to amuse the Reader: The third is by such voluminous Writings they have the more room and Liberty to Confound their Operations, speaking of one thing where they ought to speak of another; by which Preposterous manner of Writing the Searcher can't fail of being bewildered; this is not a Fault about the Operations only, but also about the time of the Operation, which causes Abundance of difficult Thoughts in the Operator, and makes him many times think those things concerning Time, which are neither Probable in Nature, nor possible for Art to perform: For what is to be gathered from the most concurring Writings of them all is, that the Matter can't pass the first Dissolution in less than five Month and the riper and higher Matters are carried, the sooner and shorter will an Operation be, for in Conclusion it may be brought to the Work of a Months, then of a Week, and lastly of a day, which Operations being misplaced cause this Error; for there is great difference between that Operation, where there is ripe ferment and that where there is not; for 'tis very difficult to bring bread to rise without Yeast.

Now the last and most Principal difficulty of all is, the want of Substance or Money of your own to carry on your search and Labours; for though you have never so much Knowledge and have overcome all the other difficulties, yet without money to build Furnaces, buy Glasses and Convenient vessels, and

Coals, you can't go forward with your Operations; I make no doubt but this was the state of the Cleine Boer, and of that Worthy and famous Count Bernard Earl of Trevisan, to detain him three years from the Possession of the Magistery after he had the true Knowledge of it, and it hath been the Case of many a Worthy Artist, I am sure it hath been very often mine, which is the most difficult and deplorable Case of all, having a large Family and their subsisting while you are in search, for it requires the whole man, and so takes him from all other business, and if he makes a Friend, he is obtained with the greatest of all difficulties; for you must first discover your Subject; secondly your Operations; Thirdly the Time, as to the two former, let him be never so ignorant, he must be the Judge, and if he does not like it, you then loose both your Friend and your Art to boot; and indeed tis very difficult when a man goes about such a thing to know who is Qualified for it, or what use he will make of it when obtained:

And as to the time, he is very nice in it, if he do except of your Proposals, and to have an Operation performed to every Punctilio; and if it is not you must expect Reprimands, and sometimes the loss of your Friend; who, lying as it were, on a Bed of ease, little knows the hardship, Fatigues, Labours, Losses and Disappointment, which the Artist sustained and is subject to; neither indeed dares he to open the same for fear left they should become his Enemies; these things being rightly Considered may be reason sufficient to deter many a Worthy Labourer from his Search, to the great loss and Detriment of Art: That none of these, or such like

difficulties may be thy Portion, O Reader! is the
desire of him, who shall Conclude this Chapter with
his Well-wishes to every Sincere Searcher after Art.

CHAPTER V

THE USE OF THE LIQUOR ALKAHEST, CIRCULATUM MINUS

OR GREAT HILECH OF BELMONT AND PARACELSUS.

In this Chapter, I shall come to shew you the Use of
the Circulatum Minus, Liquor Alchahest, or Sal
Circulatum of Belmont and Paracelsus in dissolving
Universally all sublunary Concretes into their first
Matter, none excepted, for nothing opposes it, but
its Compeer or Water, and the Central heart, of
Mercury; the one destroys, it, and the other remains
untouched by its Activity; for all other Beings are
so Essentially dissolved, as that they may be
brought over the Helm, in their true Essences; nay
by Cohobation they may be reduced to an Elementary
Water; therefore for the good and Benefit of
Mankind, I could be heartily glad that the
excellency and Utility of this Menstruum, were
better known, since Helmont, Paracelsus and Starkey,
put such Noble Encomiums on it: for as the first of
these says, In Nature there is but one Fire, which
is our consuming Vulcan, none other partaking in
Virtue and Quality with it, all the true Adepti have
an undeniable Proof of, which indeed is far more
powerful than any Common Fire, for what will remain
there, as unconquerable, is by this Liquor destroyed
and Altered radically and fundamentally; the
Mechanical Practice with it is thus.

Let this Liquor or Fire be distilled from any Metal
soft and Imperfect, and it doth at first or second
time leave them in a fusible Substance like Wax, of

which the Sulphur or Tincture is dissolvable in the best Spirit of Wine, and from the residue (being kept three days in a vaporating heat) Mercury quick and running may be separated, the same may be done in harder Metals, yea, in Perfect Metals, in a longer time, by oftner reiterated Cohobations.

But this Fire being once distilled from Mercury Vulgar, it leaves it Coagulated and Fixed, so the it will endure the Test of Saturn: It's left spungious like to a Pumice-stone but heavy like Turbith Minerale, brittle and therefore without difficulty Pulverisable, which then Cohobated with Water, distilled from Whites of Eggs it causes that distilled Water to stink, but becomes of the Colour of the best Coral, whence its denominated Arcanum Coralinum.

This Fire if it be distilled from any Gem or Stone subtly Pulverised, it turns into a meer Salt of equal Weight to the Gem or Stone; Pearls it resolves into a Milk, which is their first Ens; so also Crabs Eyes (as they are Vulgarly called, being otherwise no Eyes, but Stones found in the head of the Crab,) and all Vegetable Stones, as Peach-stones, Date-stones, or the like.

In a Word, this Fire or Liquor resolves all Vegetables, Animals and Minerals, into their first Ens, and in such Concretes as Contain in them Heterogenities, it doth discover and sever (that it makes separable) the same.

But observe, this dissolution is not performed like that, which is made with the Mercury of Philosophers for that dissolves Bodies by way of Generation, but this by way of Destruction, in that it seperates between the Central Mercury, and Sulphur of Bodies, and although they are very prevalent as to Medicines, yet are they totally bereaved and divested from any generative Power; so that 'twill be in vain for any to expect, Generation therefrom, seeing the Liquor itself is prepared by the way of Wrath; and so it dissolves Bodies; therefore called Ignis Gehennae; the Fire of Hell; but the Medicines prepared by it surpassing others, I shall give you some Particular Examples of it, first of such, as are of an Inferior Rank, as to Preparation, and then of those more difficult and Noble.

Now for small Experiments and for the more ready use of the Alchahest, 'tis good to provide yourself of Convenient Vessels, as small as Egg-glasses, think and strong, with short Necks, wide mouths, and Ground stoppers exactly fitted; also small Retorts with Ground stoppers, which may serve both for digestion and Distillation; but for great Experiments and larger quantities I advise you to use my hard Metal Jugs made sizable with very long Necks, well tryed which is by putting them into a Pail or Tub of Water within two Inches up to the mouth then blow your Breath, if there be any Air hole, the Water will bubble, then not fit for use; These serve for Digestion, Dissolution, and also Distillation, because you may work them either standing upright, or lying down, as your occasion shall require; being thus provided with Vessels you

135

may begin your Solutions first on Vegetables; which
it does Resolve into their first liquid Matter,
distinguishing in them all the Heterogenities by
several Colours, and distinct places, one above
another; in which Resolution there always seats
itself in a Distinct place a small Liquor, Eminently
distinguishable from the rest in Colour, in which
the Crasis of the whole Hearb, Tree or Seed, doth
reside: in which Retrogradation of the Concrete, by
this way of Dissolution, there is no less of Vertue,
but an Exalting of the same by many degrees only
whatever virulency is in the Crude Concrete, by this
Operation is wholly extinct, with a Preservation
notwithstanding of all the specifick Vertue,
appertaining to the Concrete in its simplicity.

And furthermore 'tis to be observed, you may
dissolve all Herbs into their Principles, liquid
without Sediment, of which part will be unctuous and
fat, especially in Trees, Gums, Seeds and many
Roots; and part Aqueous in which the Volatile Salt
of the Concrete will appear to the taste, the Liquor
with its own Oyl you may Circulate into an Essential
Salt, which is indeed the first Ens of the Concrete;
but if you would have things done in a lesser time,
make your Dissolutions in a stronger heat, and
distill over your Liquor with the dissolved Body in
a due fire, so will the Oyliness be wholly turned
into a Saline Spirit, which in a distillation by
Bath will come over in various Colours the Crasis
separating itself from the Flegme (both by Colour,
Taste and Smell, as also by its Time of Coming over
the Helm distinguishable) and your Liquor left
behind at bottom, as much in quantity, and as
Effectual in virtue, as before; as for example, This

Work does happily succeed with Bawm, or any other
vegetable which is better dry than Green) which
being only Macerated some hours in a gentle warmth,
you will see it so dissolved in such a wonderful
manner, that you cannot sufficiently admire the
Effect; the Alkahest being separated from it (or
brought over according to the former direction) out
of Bawm you have a Noble Cordial for the heart: and
thus out of Helebore you may obtain a Noble
specifick against the Gout, Hypochondriac Melancholy
Calenture, and Deleria's in Fevers : out of
Colocynthida an excellent Febrifuge; and out of
Caedar an Ens for long life: For which take the Wood
Caedar ℥ iiij of the Dissolvent an equal Proportion
and digest twenty four Hours, and it will be wholly
dissolved in the Conservation of even the very same
odour, the Liquor being separated, it will freely
dissolve in Spirit of Wine, or if you first dissolve
them in Spirit of Wine, the Liquor will dissolve
with it; digest and draw off the Spirit of Wine, and
then you may distil off the Alkahest with the
Essence of the Concrete, and separate them, as you
have been directed: Observe the Dissolution of this
for long Life, must be in a gentle heat like that of
the Sun in the Spring and after that digested in a
like heat till the Oyl and Water be united into an
Essential Salt: I should advise all Vegetables to be
prepared in the like Nature, if you desire, to have
their Eminent Vertue, without losing those peculiar
Excellencies, which depend on the Vita Ultima of the
Concrete, otherwise a speedier Preparation makes the
Medicine no less Effectual for Curing Diseases,
though less powerful as to long Life: out of Myrrh,
Aloes and Safron, an Excellent Antihectical
Medicine, as also against Lypothymy's Deliquia's,
Convulsions, and Palsies. Thus much for Vegetables,

I shall now give you a short Survey of Stones, Pearl and Coral, and lastly of Minerals: Though I must confess by the way, that if your Liquor does radically dissolve a charcoal, it is as certain a sign, that it is true, as if it did dissolve Gold itself; for according to Belmont, the Work succeeds well upon Charcoal; but 'tis admirable to see how the Operations will be changed and varied according to the Degree of Fire, and diuturnity of digestion.

Take of the Stone Ludus, in Subtil Powder, and of the Dissolvent ana, Q V, digest twenty four hours then distill, and 'twill be converted wholly into a Salt, which being Calcined will in a cold Moist Air easily run p. deliq: which will certainly Cure the Stone with all its Attendants.

Take of Pearl, what quantity you please; and of the Liquor, equal Proportion, which being therein Immerged, 'twill dissolve into a Mucilage (a gentle Maceration of some hours preceding) Resolvable in Spirits of Wine. The same may be done on Crabs Eyes, but sooner: 'Tis an excellent Medicine, for comforting the heart, giving strength to the very Marrow and Bones: Coral so dissolved is a Medicine that restores sense to those bereaved of it, Comforts the Brain, Memory and Heart, expelling sadness and Melancholly, and making a chearful and healthy Constitution.

Observe with this Liquor you must use no acid Spirit, or Salt, or Corrosive of what sort soever. For where ever such things are used, as Mediums whether for Mercury, or any other, they must be well washed off and made sweet before the Alkahest is put on; therefore in Sulphur fine Flowers are the best: Of these take what quantity you please, of Liquor Equal Proportion Digest for the space of two days, and afterwards Cohobate twice or thrice, they will come over the Helm in form of a very Red Oyl, separable from the Liquor by a separating Glass: Excellent in the Consumption, Coughs , and the like; us not only a Preservative of Man's body but also of Beer, Wines and other Liquors.

If you Abstract this Liquor from the Calx of Lead, twenty four hours digestion being premised; you will have the Lead so Reserated unlocked or opened, as that in Spirit of Wine, 'twill easily let go its Sanguineous and sweet Tincture: which is the Magistery of Lead, and an Excellent Medicine for all burnings and Inflammations whatsoever.

Take of the Flowers of Antimony, sublimed with Sal Ammoniack and dulcified, or of the Alcohol of Antimony, which is better, one Ounce of the Liquor Alkahest, three Ounces, put them into a Retort, and digest six hours, at furthest then still off the Dissolvent, and you will have a true Medicine, which Infallibly Cures the Dropsie.

Take of Precipitate very well edulcorated, made after what manner you please, one Ounce, of the Dissolvent two Ounces, and having been digested 24 hours Distil, and you'll have a fixed Pricipitate, working by stool, sweat and Urine, a certain Remedy for the Leprosy, Scurvy, King's Evil, Gout, and Pox.

Take of the Calx of Gold one Ounce, of the Liquor two Ounces; digest in a Viol with a long Neck (or one of the Egg-glasses, before described which is better) for three days, or until it will give forth no more Tincture; then pour off all that is dissolved into a Retort, and with a gentle Fire draw off the Liquor, and you'll find the Gold dissolved in the Bottom of the Retort, which you may either dissolve in Spirit of Wine, or let run in the Air per. deliq.; and you have a true Aurum Potabile: The same Process is to be observed in Silver.

Another; Take Gold Calcined into fine Atoms, or Laminated into thin Leaves, one Ounce, of the Liquor Alkahest, three Ounces put them into a Retort with a ground stopper, and let them remain in the heat of a gentle Bath a few daies, or until the Gold be dissolves without Sediment, the Liquor then being distilled from it, leaves it in the form of a Salt fusible which Cohobated often with the Liquor, is made Volatile, and comes over in two Liquors, White and Red; the Red is the Hematine Tincture, and the White may be reduced into a White Mercurial Body, after the dissolving Liquor is separated from the same: Thus Gold the King of Metals, of Nature most fixed in Corrosives, Test and Cuppel enduring all kind of Martyrdomes without the least diminution, even the most exquisite Tryal of Vulcan is by this Liquor or Fire, wholly mastered and Conquered so as to be brought into its Mineral Ens, which is the highest Preparation of Gold, that can be made by

means of this Liquor, being its Fifth Essence, and
is of Power to cure the most deplorable Diseases, to
which the Nature of Man is subject; but the
Magistery of Gold, which is the first Preparation of
it, by means of this Liquor, is a most eminent
Medicine against all Malignant Feavers the

Pestilence, Palsies, the Plague, and ℀. In the like
manner you Prepare the Fifth Essence of Silver; but
this following Medicine, is equal if not superior to
either.

The Sweet Oyl of Venus Take of the best Danzick or
Romin Vitrol, and Calcine it till it be thoroughly
wasted in the Fire what will wast; then dulcifie the
Colcothar with distilled Rain Water and dry it very
well, to the Vitriol thus prepared, add of the Fire
or Liquor , equal Parts for it will be dissolved
easily and Friendly, distil off your Liquor , and
pour it back again; and thus Cohobate it at the
least 12 or 15 times, so will all the Body of the
Colcothar be brought over the Helm, in form of a
Green Liquor; digest this same in the gentle heat of
a Bath, for about a Month and then distil it in a
slow Fire, so will the whole Metalline substance of
the Venus, come over, leaving the Liquor below in
the Retort, in its entire Pondus and Virtue: To this
Liquor or Spirit put an equal quantity of Sal
Armoniack, dissolved in as much Water as will
dissolve it, so shall you separate the Green Liquor
from a White Sediment, which White Sediment, will
give a White Metal, as fixed as Silver and will
abide the Test of Saturn, but yet formally distinct
from Silver which thou (if a Philosopher) shalt
easily perceive however as good to a Metallurgist as
the best Silver; the green Liquor dry up in a Viol

141

Glass, by evaporating all the Moysture, for its the Sulphur of Venus mixed with the Sal Armoniack, by which (Note that) it is fixed so that it will abide all Fire, this Sulphur extract with the most pure Spirit of Wine, which will dissolve it, leaving the Sal Armoniack; distil away then from it (thus dissolved) your Spirit of Wine, and you have left a very fragrant green Oyl of Venus, which is Sulphur Essensisicated, by these Operations, as sweet to taste as the best Honey, than which Nature hath not a more Soveraign Remedy for most (not to say all) Diseases: this is the true Nepenthe of Philosophers, causing certain Rest, and asswaging all Paines, but ever after sleep leaving the Party either sensibly amended (in more violent and diurnal Diseases) or quite well in the less rigid Maladies: Thus also from Lapis Hematitis and Spelter may be had Noble Medicaments, also from the Sulphur of Antimony, and more especially from Common Mercury; for if you Cohobate the Liquor so long till its body is brought over and proceed in all things as in the Sulphur of Venus, you have a Medicine that will effect whatever can be desired by either Patient or Doctor.

Thus having given you a short Landskip (as it were) of this Liquor, I shall here pass it by at present, and Conclude.

FINIS

Aphorisms of Urbigerus

Aphorismi Urbigerani, Or Certain Rules, Clearly demonstrating the Three Infallible Ways of Preparing the Grand Elixir or Circulatum majus of the Philosophers.

London, 1690

To our Dear Disciples, Honor'd Coadepts and all Well-Wishers to our Hermetic Art.

Finding you, dear Sons who have through our means attained to the true knowledge of our first Matter, worthy to receive our farther Instructions in the remainder of the Process, to extirpate all such Ambiguities, as you may have conceived in our Absence, to facilitate your Labors, and to precaution you in the bringing your Work to its Highest Perfection. We here, according to your desire, expose to you, and for your sakes, to the Public, all the most infallible Rules, necessary for preventing of Errors in this great Undertaking And tho you, ever - honored Coadepts, could never yet so far prevail on yourselves as to come to a Resolution of presenting the World with the full Practice of this our Art, joined to the Theory, we are nevertheless most certain, that we shall not receive any Reprimand from you for bringing to light these our Rules which we have so penned, that even those, who know not our Person, will not only soon perceive, that all we have written, is the real Truth, clearly exhibiting both the Theory and Practice of the whole Hermetic Art, but also conclude, that these Operations must of necessity have very often passed through our own hands from our giving such positive Rules and infallible Instructions, elucidating all the most obscure and intricate Enigrns of time Philosophers, and warning them of all the Accidents, that may happen is the working of our Subject. We are, we say, confident, you neither will, nor can blame us for this: since you will easily discern, that our Design is purely to instruct our Disciples, and prevent all the Well-wishers to this most noble Art from being imposed upon, and cheated, by any false pretended Adept: to the end that those, who shall from the Divine Benignity, by the help of these our Aphorisms, or otherwise, have received the blessed knowledge of our first Matter, which is the very same in all our three ways of producing the grand Elixir, may through these our certain Rules obtain the accomplishment of their Desires.

Having in our Travels fortuned to meet with some Persons of true Principles in Philosophy and Religion, we could not but embrace them and instruct them towards its farther Perfection, which cannot be attained without the true knowledge of our Celestial Art, by which comprehending all the Mystery of Mysteries, we learn also how to serve.

God in Faith and Truth.

And since we have no Obligation to any living Soul for time knowledge, we possess, having attained it all by the only Blessing of Almighty God on our Industry and Experiences being therefore at more liberty than those, who receive such a Favor from us, or some other Adept, 'tis our Determination, whenever we meet with Persons so qualified, always to do the same. Wherefore being at present in England though we are no Native of this Kingdom we think it necessary to set forth these our Aphorisms in the English tongue not in the least doubting, but that the Knowing, minding only the sense, will easily pardon any Impropriety, they may find in our Expressions: and when Providence shall carry us into any other Country, we, having attained to some competent Knowledge of most European Languages, shall again take care to publish them in the Speech

of the Place, where we shall be. that we may the
sooner obtain the effects of our Desires, which aim
at nothing, but the undeceiving of the World by
setting down certain and evident Marks,
distinguishing the Worthy from the Unworthy, and at
the bringing of Men to leave their unnecessary
Forms, by instructing them in the true way of
Serving God, being the only means to render them
happy both in this World, and the next.

Aphorismi Urbigerani

Or
Certain Rules, clearly demonstrating the Three infallible Ways of preparing the Grand Elixir of the Philosophers.

I. The Hermetic Science consists only in the right knowledge of the first Matter of the Philosophers which is in the Mineral kingdom not yet determined by Nature.

II. An undetermined Matter being the beginning of all Metals and Minerals, it follows, that as soon as any one shall be so happy, as to know and conceive it, he shall easily comprehend also their Natures, Qualities, and Properties.

III. Although some Persons, possessed with foolish Notions, dream, that the first Matter is to be found only in some particular places, at such and such times of the year, and by the Virtue of a Magical Magnet; yet we are most certain according to our Divine Master Hermes) that, all these Suppositions being false, it is to be found everywhere, at all times and only by our Science.

IV. The Hermetic Art consists in the true Manipulation of our undetermined Subject, which before it can be brought to the highest degree of Perfection, must of necessity undergo all our Chymical Operations.

V. Our Chymical Operations are these Amalgamation Sublimation, Dissolution Filtration, Cohobation,

Distillation, Separation, Reverberation, Imbibition, and Digestion.

VI. When we call all these Operations ours, they are not all to be understood according to the common Operations of the Sophisters of Metals, whose Industry consists only in disguising of Subjects from their Form, and their Nature: but ours are really to transfigure our Subject, yet conserving its Nature, Quality, and Property.

VII. This our Subject, after its having passed through all those artificial Operations, which always imitate Nature, is called the Philosophers Stone, or the fifth Essence of Metals, being compounded of the Essence of their four Elements.

VIII. Time Metals and Minerals, which Nature already determined, although they should be retrograded into running Mercury Water, and Vapor; yet can they by no means be taken for the first Matter of the Philosophers.

IX. Our true and real Matter is only a Vapor, impregnated with the Metallic Seed, yet undetermined, created by God Almighty, generated by the Concurrence and Influence of time Astrums, contained in the Bowels of the Earth, as the Matrix of all created things.

X. This our Matter is called undetermined, because, being a Medium between a Metal and a Mineral, and being neither of them, it has in it power to produce both, according to the Subject, it meets withal.

XI. Such a Metallic Vapor, congealed and nourished in the Bowels of the Earth, is called the undermined, and when it enchants the Serpent with the Beauty of its termined and additional Fire, the determined Green — Dragon of the Philosopher and without the true knowledge and right Manipulation of it nothing can be done in our Art.

XII. This Green—Dragon is the natural Gold of the Philosophers, exceedingly different from the vulgar, which is corporeal and dead, being come to the period of its Perfection according to Nature, and therefore incapable of generating, unless it be first generated itself by our Mercurial Water but ours is spiritual, and living, having the generative Faculty in itself, and in its own Nature, and having received the Masculine Quality from the Creator of all things.

XIII. Our Gold is called Natural, because it is not to be made by Art, and since it is known to none, but the true Disciples of Hermes, who understand how to separate it from its original Lump, 'tis called also Philosophical; and if God had not been so gracious, as to create this first Chaos to our hand, all our Skill and Art in the Construction of the great Elixir would be in vain.

XIV. Out of this our Gold, or undetermined Green—Dragon, without the addition of any other created thing whatsoever, we know how through our Universal Menstruum to extract all our Elements, or Principles, necessary for the performance of our great Work: Which is Our first way of preparing the Grand Elixir and since this our first Chaos is to be had without any Expense, as costing only the trouble

of digging it out of the Mines, This is not unfitly
called the only way of the Poor.

XV. The Operations in this our first way being in a
manner the same with those of our second, which is,
when we join our determined Dragon with our Serpent,
we shall (to avoid Repetitions) in the subsequent
Aphorisms give Instructions for them both together.

XVI. Our Serpent, which is also contained in the
Bowels of the Earth, being of all created things
whatsoever the nearest subject of a Feminine Nature
to our Dragon, through their Copulation such an
astral and metallic Seed, containing our Elements,
is also to be brought forth as can, though with
somewhat more of Expense and Time, perform the whole
Mystery of Hermes.

XVII. Since our Serpent is of all created things the
nearest subject of Feminine Nature to our Dragon,
she is after her Copulation to be taken for the
Basis of our Philosophical Work: for out of her
Bowels, without the help of any other Metal or
Mineral, we must draw our Principles or Elements,
necessary to our Work, being retrograded by the
Universal Menstruum.

XVIII. This Feminine Subject cannot be retrograded,
unless to free her from her Impurities, and
Heterogeneous Qualities, she is first actuated by
her Homogeneous ones, that she may be in better
Capacity to receive the spiritual Love of our Green
Dragon.

XIX. After our Serpent has been bound with her
Chain, penetrated with the Blood of our Green

Dragon, and driven nine, or ten times through the combustible Fire into the elementary Air, if you do not find her to be exceeding furious, and extremely penetrating, 'tis a sign, that you do not hit our Subject, the Notion of the Homogenea, or their Proportion.

XX. If this furious Serpent, after it has been dissolved by the Universal Menstruum, filtrated, evaporated, and congealed nine or ten times, does not come over in a Cloud, and turn into our Virgin Milk, or Metallic argentin Water, not corrosive at all, and yet insensibly, and invisibly devouring everything, that comes near it, 'tis plainly to be seen that you err in the Notion of our Universal Menstruum.

XXI. The Serpent, of which I now speak, is our true Water of the Clouds, or the real Eagle and Mercury of the Philosophers, greatly different from the Vulgar, which is corporeal, gross, dead, and full of Heterogeneous Qualities, and a Subject fallen from its Sphere, like unripe Fruit from the Tree; but ours is spiritual, transparent, living, residing in its own Sphere, like a King on his Throne.

XXII. Though the vulgar Mercury is such an unripe fruit, corporeal, and dead; yea, if you know how to amalgamate it with our Dragon, and to retrogradate it with the Universal Menstruum, you may assure yourself, that out of this also you shall be able to prepare a Sophic Mercury, with which you shall certainly produce the great Elixir, discover the Secret of Secrets, unlock the most difficult Locks, and command all the Treasures in World.

XXIII. Our Mercury is called the Mercury of the Philosophers, because it is a Subject, which is not to be found ready prepared to our hand : for it must of necessity be made by our Philosophical Preparations, out of the first Chaos, and although it is Artificial, yet it is naturally prepared, Nature, which is imitated in the Preparation of it, contributing likewise thereunto.

XXIV. Since our Subject cannot be called the fiery Serpent of the Philosophers, nor have the power of overcoming any created thing, before it has received such Virtue and Quality from our Green—Dragon, and the Universal Menstruum, by which itself is first overcome, devoured, and buried in their Bowels, out of which being born again, 'tis made capable of the same, it follows, that such a Virtue of killing and vivifying is natural to our Dragon and the Universal Menstruum.

XXV. The Universal Menstruum of the Philosophers is that Celestial one, without which nothing can live nor subsist in this World: 'tis also that noble Champion, which delivers time uncorrupted Virgin, Andromeda, who was with a strong Chain fastened to the: Rock in the power of the Dragon, of whole spiritual Love having admitted, for fear of being eternally ruined and devoured by him (which could not have been avoided, if this noble Champion had not come to her assistance) She is to be delivered of a Child, which wilt be the Wonder of Wonders, and Prodigy of Nature.

XXVI. If our Virgin in her Confinement, before she is set at liberty, does not manifest her extreme Beauty with all her internal, divers, delicate natural Colors wonderfully charming, and very pleasant to the Eye, it signifies, that she has not sufficiently enjoyed the spiritual Company of the Dragon.

XXVII. If the Universal Menstruum has not totally delivered the Virgin from the Claws of the Dragon, it is a sign, either that she was not sufficiently free from her Heterogeneous Qualities, or that she had not received from the external Heat a sufficient penetrating Quality, or that the Universal Menstruum was too weak to perform its Undertaking.

XXVIII. To know, whether the Amalgamation, Sublimation, Dissolution, Filtration, Coagulation, and Distillation have been Natural and Philosophical, the whole Body of the Serpent must come over spiritual and transparent, leaving only some few and very light Feces at the bottom, which can by no Art be reduced! — either into a running Mercury, or another kind of metallic Substance.

XXIX. After all these above—mentioned Operations, and the Separation, if our Serpent, being amalgamated with any Metal, pure or impure, cannot suffer the Fusion, it will be in vain for you to go any farther with it: for you may assure yourself, that you do not walk in the true Paths of the Hermetic Art.

XXX. Our Philosophical Distillations consist only in the right Separation of our Spiritual and Mercurial Water from all its poisonous oily Substance, which is of no use at all in our Art, and from though Caput Mortuum, which is left behind after the first Distillation.

XXXI. If after the first Distillation an exceedingly corrosive and extremely penetrating red Oil does not ascend (which as soon as it begins to appear in the Neck of the Retort, the Receiver must be changed it signifies that the Distillation has not been rightly performed, and by Consequence, that the internal Fire of our metallic vaporous Water, being burnt up, and corroded by its poisonous Vapor, and the outward Fire, is still mixed with it, and with the Caput Mortuum.

XXXII. In case you should commit so great an Error in the performance of this first Distillation, although it will never be in your power to prepare the Mercury Duplex of the Philosophers, unless you should begin the whole World agar; from the very beginning; yet, if you have any farther Skill in our Art, you may easily prepare our Mercury Simplex, with which you will effect great and miraculous things.

XXXIII. This blood red Oil with its only Fumes penetrates every Part and Atom of all Metals and Minerals, and principally of Gold out of which Dissolution one may easily extract the right Tincture or Essence with highly rectified Spirit of Wine, and bring it over the Alembic with it: which is indeed a great Medicine for humane Bodies.

XXXIV. A deep blood—red Tincture of excellent Virtue is to be extracted also out of the above—mentioned Caput Mortuum, accidentally and unfortunately intermixed with the internal Sulfur of our Mercurial Water, and with the red Oil, with highly rectified

Spirit of Wine : with which after it has been evaporated to a Powder, imbibed, and Philosophically digested, you may assure yourself of having the Medicine of Medicines, next to the great Elixir, by which you may, imperceptibly and quickly cure all sorts of Distempers, to the great Admiration. Of all Galenists, and to the Astonishment of all Vulgar Chymists.

XXXV. The most part of the Philosophers, whilst their Intention was to go farther to the noblest Perfection of our Celestial Art, either employed this red Oil brought to a Potability, for internal Medicines, or to external Diseases without any farther Preparation of it, till they had obtained the great Elixir.

XXXVI. If the Caput Mortuum has not the Magnetic Quality in attracting the Spiritus Mundi into itself from the Astrums, it is a sign, that at time end of the Distillation of the red Oil the outward fire was so violent, as quite to burn up the Magnet, which is contained in the first Feces of our Mercurial Water.

XXXVII. After the first Distillation, if the least Part of the Virgin Mercurial Water can by any Art whatsoever be brought to running Mercury, or any other kind of a Metallic Substance, it is an evident sign, that either the Subject, or its Preparation and Reduction into Water, has not been real, natural, or Philosophical.

XXXVIII. The above—mentioned Spiritus Mundi, although of no use at all in this our great Work, is yet a great Menstruum in extracting of Tinctures out of Metals, Minerals, Animals, and Vegetables, and in performing great things in the Art volatilizing all fixed Bodies, and principally Gold.

XXXIX. A great many Pretenders to the true Hermetic Knowledge prepare Menstruums, to dissolve common Mercury, and to turn it into Water several manner of ways, and by several additions of Salts, Sulfurs, Metals, and Minerals. but, since all those Preparations are sophistical, any one, expert in our Art, will be able to reduce it to its running Quality again.

XL. The Quality of our Mercurial Water: being to volatilize all fixed Bodies, and to fix all those, that are volatile, fixing Itself with those, that are fixed, according to the Proportion of it, dissolving its own Body, it unites inseparably with it, conserving always its own Qualities and Properties, and receives no Augmentation from any other created thing, but only from its crude Body.

XLI. Our Mercurial Water has such a sympathy with the Astrums, that, if it is not kept very close, and Hermetically sealed, it will in a very short time, like a winged Serpent, fly away in a wonderful manner to its own Sphere, carrying along with it all the Elements and Principles of Metals, and not leaving so much as one single drop, or the least remainder, behind.

XLII. Several Pretenders to the Magical Science prepare Magical Magnets, to draw from the Air, and (as they pretend) from the Astrums such Menstruums, as they think necessary for she Production of the Great Elixir; but their Magnets being compounded of several determinate things, although their

Menstruums are great Dissolvents yet we do on assured knowledge affirm that they can never perform any real Experiments in our Art.

XLIII. Some are of Opinion, that, unless the Operator is Master in the Magical Science, and fundamentally understands all its Experiments, he will never be able by any other Art whatsoever to bring forth any such things, as can produce the Universal Elixir. Now, although we do not deny, that the Magical Knowledge is required to attain to the highest degree of Perfection in all Sciences, yea; we are most certain, that it is not at all necessary to the Formation of the Grand Elixir upon Animals, Metals, Precious Stones, and Vegetables.

XLIV. Our Virgin Milk, or Metallic Water, being brought to a perfect Spirituality, and excellent Diaphanity, is called the true Chaos of the Philosophers: for out of that alone, without any addition of any created, or artificially prepared thing, we are to prepare and separate all the Elements, which are required to the Formation of our Philosophical Microcosm.

XLV. To understand aright, how out of this our Chaos we are to form our Philosophical Microcosm, we must first of necessity rightly comprehend the great Mystery and Proceeding in the Creation of the Macrocosm: it being extremely necessary to imitate and use the very same Method in the Creation of our little one, that the Creator of all things has used in the Formation of the great One.

XLVI. When our Chaos or Celestial Water has purified itself from its own gross and palpable Body, it is called the Heaven of the Philosophers, and the palpable Body the Earth, which is void, empty, and dark: And if our Divine Spirit, which is carried upon the face of the Waters, did not bring forth out of the palpable Body that precious Metallic Seed, we should never be able by any Art whatsoever to go on any farther with time perfect Creation of our Microcosm according to our Intent.

XLVII. This Heaven of the Philosophers, after it has Separated itself from the Earth, containing our Philosophical Seed, and the Magnet of our Salt of Nature, and from the superfluous Waters, is called the Mercury simplex of the wise men, for whosoever attains it, at the same time attains also the Knowledge and Power of retrograding Metals, Minerals, &c. so as to reduce them to their first Being, to perfect imperfect Bodies, and to vivify dead ones, conserving always its own Property and Quality to itself, and to produce the Great Elixir according to the usual ways of the Philosophers.

XLVIII. After we have separated the Water from the Water, by which I mean the Mercurial Celestial Water from the superfluous Water, which is the Flegm, by the Blessing of God and the Infusion of our holy Spirit, we do not in the least doubt, but we shall be able to bring forth out of our Earth such Fruits and Subjects, with which we shall certainly perform the whole Creation, carrying our Work to the highest Degree of Perfection.

XLIX. Our Mercurial Water being of the: same Brightness with the Heavens, and our palpable gross Body, which did separate itself from our Celestial

Water, having the same Properties and Quality with the Earth, none, but Ignorance, will deny them to be the right Heaven and true Earth of the Philosophers.

L. If, after the Separation of the Spirit from the superfluous Waters, the World, in which it is contained, does not appear mighty clear, and full of light, and of the same brightness with our Celestial Water, it is a sign, that the Separation is not fully performed, the Spirit being still intermixed with the Waters.

LI. If in the space of nine or ten Weeks, or two Philosophical Months at longest, our Mercurial Water has not done separating itself from all its own Earth containing the Metallic Seed, it is an evident sign, that you have either erred in the working of it, or that its Digestion, having been too violent, has confounded and burnt up the principal Subject of the Creation.

LII. This Philosophic Earth, containing our principal Subject, after it has been separated from all the Waters, is very gently to be dried by some external Heat, to free it from its extraneous Humidity, that it may be in a proper Capacity to receive the Celestial Moisture of our Argentin Water, to which it unites its most noble Fruits, with which our Philosophical Microcosm is generated, nourished, and saturated.

LIII. If the Earth, after it has been reverberated, humected with our Celestial Moisture, does not presently enrich our Air with the divine expected Fruits, you must certainly believe, that in the drying of it the external Heat has been so violent, as to burn up the internal Head and Nature of the Earth, and consequently spoil your Undertaking as to the performance of the whole Mystery of the Creation, according to the noblest, richest, shortest, most natural, and secretest ways of the Philosophers.

LIV. In case the Earth should be totally destroyed by the violent external Heat, although it is most certain, you cannot carry on our noble Creation any farther with it; yet if you know how to amalgamate our Mercury simplex with your common Gold, which is dissolved, vivified, and renewed by it, you may be sure of effecting the Great Elixir, although neither so quick. So natural, nor so rich, as you might have done without it. And this is our third way.

LV. The Amalgamation of our Mercury simplix with common Gold consists only in the right Proportion, and in the indissoluble Union of both, which is done without any external Heat in a very short time, without which exact Proportion and right Union nothing of any Moment is to be expected from their Marriage.

LVI. Know then, that this right Proportion is ten parts of our Mercury simplex to one of your finest common Gold in filings, which is dissolved in it, like Ice in common Water, after an imperceptible manner, and as soon as the Dissolution is over, the Coagulation and Putrefaction presently follow, which Effect, it you find not, 'tis a sign, that the Mercury exceeds its due Proportion. Now when your Gold has been thus well amalgamated, united, putrefied, and inseparably digested with our Mercury simplex, you will then have only our Philosophical Sulfur, in which time one might easily have performed the whole Work, working without common: Gold.

LVII. Although our Mercury simplex is exceedingly spiritual and volatile, yet since it is the right Agent, digesting the Seed or Essence of all Metals and Minerals it will, though undigested, naturally adhere to any of them although corporeal, that shall come near it, and will never leave it, unless it be forced away by the Test, though kept in a great Fusion for many hours.

LVIII. This Mercury simplex, which before its retrogradation was of a Feminine Nature, and before it left all its own Earth, was Hermaphroditic, being powerful in both Sexes, is now become of a Feminine Quality again, and although it has lost the Masculine visible Fire, yet it has conserved its own, which is invisible to us, and with which it performs visible Operations in digesting of imperfect Metals, after its Determination with any of them.

LIX. If this our Mercury (the Proportion rightly observed) should be amalgamated with any imperfect Metal, being first determined with a fixed one, it will regenerate and perfect the same, not losing the least Particle of its Virtue or Quantity: Which Metal after the digestion of a Philosophical Month will (as most Philosophers teach) be able to resist all manner of Tryals, and will be far better than any Natural one.

LX. The Determination of our Mercury simplex with
any of the fixt Bodies is to be done by dissolving a
small quantity of Filings of red or white according
to the Color and Quality of the Metal, that you
desire to meliorate, and if you do not err in the
Separation and Union of the Subjects, you may assure
yourself of obtaining your desire after a
Philosophical Digestion.

LXI. To examine aright, whether the Mercury simplex
is rightly prepared, or come to its Perfection, one
only Drop, put upon a red-hot Plate of Copper, must
whiten it through and through, and must not part
with it, although brought into a great Fusion:
Which, if you find, it does not, it will be a plain
Demonstration, that either your Mercury is not well
prepared, or that it has not yet done separating
itself from its own Earth.

LXII. If your Mercury simplex, put upon its own
dryed Earth, does not presently unite with the
Essence of Metals, appearing deeper than any Blood,
and shining brighter than any Fire, which is a mark
of the Reception of its own internal Fire, and that
the Eagle has suckt the Blood of our Red Lion, it is
an evident sign, that you have erred in the
Manipulation of the Earth.

LXIII. This Mercury, thus impregnated with its
Essence, or Sulphur of Metals, is called the Mercury
duplex of the Philosophers, which is of a far
greater Quality, and Virtue than the simplex, with
whose Imbibitions in the Salt of
Nature, after its being saturated with the simplex,
the whole Mystery of the Creation of the
Philosophical Microcosm is maintained and perfected.

162

LXIV. To know, whether your Mercury duplex is Philosophically prepared, and sufficiently impregnated with its own internal Natural Fire, put one single Drop of it upon a red-hot Plate of fine Silver: and if the Silver is not by this Drop penetrated through and through with a deep-red Tincture, enduring the greatest fire of Fusion, it will signifie, that you either fail in the Preparation of it, or that you have not given it time enough to receive a full Saturation out of its own Earth.

LXV. This deep-red Tincture, extracted out of our Philosophical Earth, is called our Sulphur, our undigested, essentificated Gold, our internal elementary Fire, and our Red-Lion: for without its Help and Concurrence our Philosophical World cannot be nourished, digested, or accomplished, being the right Ground, and true Essence of the whole work of our Creation.

LXVI. When the Earth has lost its Soul, the remainder of it is the true Magnet, attracting the Salt of Nature from the combustible Fire after a violent Calcination for several hours: which Salt, after its Purification and Clarification, is called the clarifyed Earth or Salt of the Philosophers, which, Uniting itself with our single and double Mercury, after their Digestion, is called by our Master Hermes the Universal Spirit earthifyed.

LXVII. The Extraction, Purification, and Clarification of our Earth or Salt of Nature is to be performed by our Mercury simplex: which being put upon the reverberated Earth, will presently draw to itself, and unite itself with it, yet separable by gentle Distillation, after which the clarifyed Salt of the Philosophers is at hand.

LXVIII. Although we use our Mercury simplex in the Extraction of its own Soul out of its Body, and for the Clarification of the latter; yet, since it is a philosophical and perpetual Menstruum, it loses nothing of its connatural Prerogatives, nor does in the least diminish in Quantity, being our true Alkahest, as Paracelsus is pleased to call it.

LXIX. Those three Principles, or Elements of our Chaos, perfectly separated from their Impurities, and brought to their highest Perfection, are rightly called the three Herculean Works: for after the Preparation of them all the Labor, Trouble, and Danger will be past.

LXX. Some foolish Operators pretend, that our Great Elixir is to be prepared in a very easie manner, and without any trouble at all, to whom we will with our Master Hermes, briefly answer, That such Impostors neither know our Matter, nor the right Preparation of it. Yet we do not deny, but any Healthy Person, of what Age soever he may be, may understand all our Herculean Labors, necessary to the Performance of it.

LXXI. These our Operations are therefore called Herculean in respect to the rest of the Work, which is exceedingly easy, and without the least Trouble or Danger, being for that reason called Childrens Play, because a Child or a Woman, that has any Sense, may easily work it, and bring it to the highest Perfection, according to the Saying of all true Philosophers.

LXXII. Although all those above—mentioned Operations are, according to the common Opinion of the Philosophers, esteemed difficult, and dangerous; yet we can upon our Conscience assure you, that we have our self alone without the help of any Creature living prepared them all on a common Kitchen Fire, as is very well known to several Coadepts, our Friends, who could not but admire and approve of our Industry.

LXXIII. No true Adept or perfect Artist can deny, but that the whole Work of the Great Elixir may from the very beginning to the end be performed on one only Furnace, in one only sort of Vessel, and by one only Person alone, at a very small charge.

LXXIV. Some Impostors would perswade the Vulgar, that Gold, Silver, and many other Ingredients are required to the making of the Grand Elixir according to our noblest ways: which the Doctrines of all the Philosophers, and our own infallible Rules clearly shew to be false: for 'tis most certain, that we neither use any of their Ingredients, nor yet any Silver or Gold, (unless, as we have mentioned, in our third way) till we come to the Fermentation of our Elixirs.

LXXV. We do with all true Philosphers assure you, that all things, necessary for our Philosophical Work, besides the Fewel, Vessels, and some new Instruments, belonging to the Furnace, are to be purchased for less than the Expence of one single

Guinea, and that everywhere, and at all times of the year.

LXXVI. Since neither Gold nor Silver is to be used at all in the Formation and Cibation of our Philosophical Work, it follows, that the old and common saying of some Authors, viz. That without working with Gold 'tis an impossible thing to make Gold, proved to be only a false Notion of Men, who understand not our Art.

LXXVI. When our Herculean Works are brought to Perfection, which is, when our three Principles, or Elements are prepared, purifyed and perfected, unless the Philosophical and unseparable Union of them is exactly performed, the Great Mystery of our Creation is not to be expected.

LXXVIII. Our Principles or Elements being brought to a perfect and unseparable Union and Digestion, it is called the Triple Mercury of the Philosophers, which being finished, the whole Creation and Formation of our work is crowned.

LXXIX. All our Work of the Creation from its very Beginning to its perfect End may, on our certain knowledge, be perfected in less than nine Months by any skillful and careful Artist, that follows our Rules, unless some Accident should happen in the Preparation of our Herculean Works: which to prevent, we wrought them our self in an earthen Vessel, which we count far better and surer than any Glass, and which is most agreeable to the Practice of the most ancient Philosophers.

LXXX. Before you come to the Union of your Elements, your clarifyed Earth is before all things to be digested in a moderate and continual Heat of Ashes, to free it from any unnatural Moisture, that it might have attracted after its Purification, to be in a fit Capacity to receive your Mercury simplex, by which it is to be nourished in its Infancy.

LXXXI. If your clarifyed Earth, after it has been digested the space of a whole Month, does not appear exceeding dry, subtil, and frangible, it will signifie, that you have failed in the Purification or Clarification of it, or that the external Moisture, it had attracted, is not yet parted from it.

LXXXII. Take great Care, that you do not begin your Imbibitions of your Earth, before you find it to be very well purifyed, clarifyed, dryed, and brought to be very subtil, and extremely frangible: for it would be a great Detriment as well to your Work, as to your Mercury; and, although it should not spoil your Work, yet it would be to you a great loss of Time.

LXXXIII. After our clarifyed Earth had been brought to a perfect Purity, Dryness, and Frangibility, it is to be imbibed with the eighth part of our Mercury simplex, or Virgins Milk, which will in a very short time be soaked into it, as into a Sponge, which shews the hungry State of our Infant, and then the Fire is to be continued, till the Infant is hungry again.

LXXXIV. If in the space of two or three days, or four, at the farthest, the Infant does not shew itself to be extreme hungry by becoming very dry and frangible again, it will become an evident sign, that you have overcome it by your excessive feeding of it.

LXXXV. Great care is to be taken also in the feeding of the noble Infant: for if you do not well observe all our infallible Rules, you will never be able to bring it to a perfect Maturity: for in the Notion and Proportion of our Imbibitions, and the Management of them, the prosperous and unfailable End of our Work is to be expected.

LXXXVI. 'Tis always to be observed, that the Fire be very moderate, as long as you are making your Imbibitions, for fear or forcing any part of your Mercury to leave the Earth: for as a moderate Heat makes the Union between the Soul and the Body, and perfects all the Work: so on the contrary a too Violent Heat disunites and destroys all.

LXXXVII. The Infant being dry, the Imbibition is to be repeated again, and this Method is to be used, until the Matter has received its weight of the Mercury: at which time if you do not find it to flow like Wax, and be whiter than any Snow, and very fixt, you must proceed with your Imbibitions, until you perceive the same.

LXXXVIII. The Imbibitions are not to be made any oftner, than once every three or four days, in which time you will find your Matter, having soaked up all your Mercury, to be in great want of Food, which must be supplyed, until it be saturated: the Mark of which will be, when it flows like Wax again.

LXXXIX. Your Matter being brought to a perfect Fluxibility, uncomparable Whiteness, and unalterable Fixedness, know then, that you have perfectioned the white Elixir, which, being fermented with fine Silver in Filings, will be in a Capacity to transmute all inferior Metals into the finest Silver in the World.

XC. Before the white Elixir is fermented with common Silver, you may multiply it, as well in Virtue, as in Quantity, by the Continuation of Imbibitions with the Mercury simplex, by which it may by Degrees be brought ad Infinitum in its Virtue.

XCI. The white Elixir being brought to its Degree of Maturity, desiring to go on to its highest Degree of Perfection, instead of fermenting it with Silver, it must be cibated with its own Flesh and Blood, which is the double Mercury, by which it being nourished, multiplied in Quality and Quantity, and digested, the whole Work is accomplished.

XCII. As soon as the first Imbibition is made, you will see a great Alteration in your Vessel: for there will be nothing seen but a Cloud, filling the whole space of the Vessel, the fixt being in controversy with the Volatil, and the Volatil with the fixt. The Volatil is Conqueror at the beginning, but as last by its own internal Fire, conjoyned with the external, both are united, and fixed inseparably together.

XCIII. It is to be observed, that the Glass Vessel, which must be oval, with a Neck half a foot long, and very strong, be of a fit bigness, and of such Capacity, that your Matter, when it is put into the

Vessel, may take up only the third part of it, leaving the other two vacant: for, if it should be too big, it would be a great hinderance in performing the Work, and if too little, it would break into a thousand pieces.

XCIV. After you have cibated the noble Elixir with your double Mercury, before it can come to its perfect Fixedness, it must of necessity wander through all the States and Colours of Nature, by which we are to judge its Being and Temperament.

XCV. The constant and essential Colors, that appear in the Digestion of the Matter, and before it comes to a Perfection, are three, viz. Black, which signifies the Putrefaction and Conjunction of the Elements; White, which demonstrates its Purification; and Red, which demonstrates its Maturation. The rest of the Colors, that appear and disappear in the Progress of the Work, are only accidental, and unconstant.

XCVI. By every Cibation of its own Flesh and Blood, Regeneration of its Colors, and Digestion, the Infant will grow stronger and stronger, that at last being fully saturated and digested, it is called the Great Elixir of the Philosophers, with which you will be able to perform, Wonders is all the Regions, as well Animal, as Mineral, and Vegetable.

XCVII. When your Elixir is brought to a Fluxibility, and a perfect Fixedness, if you desire to make a Medicine upon Metals, you must determinate or ferment it with common Gold in Filings, in which Determination it will vitrify, and then you will have an incomparable Medicine, capable to transmute all imperfect Metals into the purst Gold, according to the Doctrine of all the Philosophers, though our self never designed anything, but an universal Remedy for the Cure of all curable Diseases, incident to Human Bodies, as is well known to our Friends, who have enjoyed the Benefit of these our Labors.

XCVIII. It is to be observed in the Fermentation, that the Elixir exceed not the Ferment in Quantity, otherwise the Sponsal Ligament of it cannot be actually performed, and when the Ferment is predominant over the Elixir, all will be presently turned into dust.

XCIX. The best Method of Fermentation is to take one part of the Elixir, and put it into the midst of ten parts of Gold in Filings, cast through Antimony, to free it from all its Impurities, and to keep it in a circulary Fire for the space of six Hours, so increasing the Fire by Degrees, that the last two hours it be in a good Fusion, and when cold, you will find all your Matter exceeding frangible, and of the Color of the Granate—Stone. C. Common Mercury, amalgamated with Lead, is counted the most proper Subject for making Projection, which being in Fusion, your fermented Matter being divided into three parts, one part of it rolled, in Wax, is to be flung upon the Amalgam: then presently cover the Crucible, and continue the Fire, until you hear the Noise of the Separation and Union: then the second and third part, as before, and being kept for two hours in a continual Fire of Fusion, let it cool by itself.

CI. Whoever shall presume to prepare the Great Elixir according to our most Secret Ways without following and observing all these our infallible Rules, will certainly find himself mightily mistaken at last, having after a great deal of Troubles, Charges, and Pains, reaped nothing but Discontent; and on the contrary they, that shall walk in our true and infallible Paths, shall with very little Trouble and Expences attain to their desired End, which we cordially wish to all those, who are sincere well—wishers to the Hermetic Philosophy.

A Postscript, containing An Explanation of the Figure, prefixt to the Aphorismi Urbigerani.

Having in our One Hundred and One Aphorisms so perspicuously laid open all the Difficulties, and so amply taught the compleat Theory and Practice of the whole Hermetic Mystery, that any ingenious Lover of Chymistry will not only be enabled to understand the most abstruse writings of the Philosophers, but also to effect any real Experiment, which is to be expected in the Progress of our Celestial Art; and yet being apt to believe, that such, as are not our Disciples, may perhaps meet with some of the Philosophical Figures, the meaning of which they may not so easily comprehend, we have judged it highly expedient, in the Front of this our little Book to place this our Figure, by which, being a perfect Compendium of all the Philosophical Emblems, the rest may be without any great difficulty understood. Now since this our Figure, mystically representing all our Subjects and Operations, cannot but admit of many and various Interpretations, all which if we should here set down, our Aphorisms (where they are already delivered, and of which this would then be a

Repetition,) would be altogether useless and insignificant: we therefore at first esteemed it very superfluous to give any farther Illustration of it. But our desire being to do all the good, we can, to the Public, we have on second Thoughts resolved with our wonted Brevity to deliver the following Explanation for the better Comprehension both of it and our Aphorisms.

The Tree is a Supporter of the Motto, 'Virtus unita fortior': which, being to be read from the side of the Serpent, representing by the Half—Moon on its Head the Planet, under whose Influence it is born, is to be referred to it according to its particular Motto, which signifies, that, if you take it alone, it can do little or Nothing in our Art, as wanting the Assistance of others. By the Green Dragon is to be understood our first undetermined Matter, comprehending all our Principles, (as is demonstrated by the Half—Moon on its Head, the Sun in its Body, and the Cross on its Tail,) and denoting by its Motto, that it can perform the whole work without being joined with any other created or artificially prepared thing: which is our first way. But this our Dragon, when copulating with our Serpent, is forced to comply with her, degrading itself from its undetermined Being for the production of our second way. Apollo with the Sun on his head, and Diana with the Half—Moon, embracing each other, shew our third way, and the Continuation of our first and second. The River, into which they descend, signifies the State, they must be reduced into, before they can be in a Capacity of being born again, and before in any of our three ways they can be brought to a perfect Spiritualization and Union. Apollo and Diana, coming out of the River in one wonderful Body, Diana having obtained all, represent our Herculean works, ready finished and the beginning of their Conjunction, and by their going to set their foot on firm ground, where she is to sow the noble Fruits for the Procreation, is to be understood the Continuation of their Conjunction, till they are fully united and perfected. In this Scheme also, as well as in our Aphorisms, are mystically exhibited all the principal Points of Faith and Religion, comprised in the Volumes of the Old and New Testament: whence it manifestly appears, that the Contemplation of Nature truly leads to the Comprehension of those heavenly Verities, by which alone we can expect to arrive at the Enjoyment of that blessed Immortality, to which, as to the true and ultimate End of our Creation, all our Endeavours are to be directed.

-Finis-

A Word from the Publisher

Thank you for purchasing this small work from The R.A.M.S. Library of Alchemy. During his lifetime, Hans Nintzel was dedicated to the identification, acquisition, study, retyping and, when necessary, translation of what he considered to be the most important known works on Alchemy. Hans was assisted by his sparse network of fellow Alchemists, all members of the Restorers of Alchemical Manuscripts Society (R.A.M.S.). I was an active member of R.A.M.S.

My goal is to publish all of the works originally made available through R.A.M.S. as photocopies. To facilitate this, I have chosen to have the books professionally printed. I also have a few titles that I intend to add to the original R.A.M.S. Library, selected by strict criteria established by Hans.

The works from the original R.A.M.S. Library are republished by R.A.M.S. Publishing Company in the collection, "The R.A.M.S. Library of Alchemy," with permission of the Estate of Hans W. Nintzel.

If you have a work on Alchemy that you believe should be a part of the R.A.M.S. Library, please contact me through R.A.M.S. Publishing Company.

Philip N. Wheeler

www.ingramcontent.com/pod-product-compliance
Lightning Source LLC
Chambersburg PA
CBHW080809180526
45168CB00006B/2376